Praise for the third edition of
Developing Ecological Consciousness

"A field guide to our deep connection to earth, a story of why we need to care, and an invitation to walk the path towards healing the planet and ourselves."
—**Doug Wentzel, senior naturalist,**
Shaver's Creek Environmental Center

"As a student of the humanities and an advocate for human rights, I've never considered myself an 'environmentalist' or 'eco'-anything. So an ecological textbook that begins with a quote from Vaclav Havel, and even makes room for poetry and religion, has my undivided attention. Indeed, if I'd read this book in college, I might have understood far earlier what that ecologist and radical abolitionist Henry Thoreau taught me years later—that we are all of us 'part and parcel of Nature.'"
—**Wen Stephenson, author of *What We're Fighting for Now Is***
Each Other: Dispatches from the Front Lines of Climate Justice

"In this thoroughly revised edition of *Developing Ecological Consciousness*, Christopher Uhl helps us answer one of the most critical questions of our time, 'What gives you hope?' While Uhl's own answers are both wise and heartfelt, the essential contribution of this book is the guidance it gives readers in developing their own answers. Viewed as a 'guidebook for hope,' this should be required reading for anyone, from university faculty with their students to parents with their children, trying to discern, or reimagine, what it could mean to be fully human in these precarious and purposeful times."
—**Jeffrey Gerwing, Portland State University**

"*Developing Ecological Consciousness* is a sober, empowering, learned, and impassioned guide that summons us back to the fullness of our shared humanity. It is an agent of activation, helping us shake off the inertia we commonly feel when pitted against the monumental scale of our ecological crisis. It helps us understand what the crisis is specifically asking of us: to disentangle ourselves from the outmoded paradigms of thinking we are schooled in, from the half-life of stale relationships, from the narrow anxieties that bind us, and grow into a fully awakened relationship with the natural world. Full of surprises, this book brilliantly illuminates what I believe to be the only way we can move forward. Read it, and feel your life being enriched."
—**Philip Shepherd, author of *Radical Wholeness***
and *New Self, New World*

DEVELOPING ECOLOGICAL CONSCIOUSNESS

Becoming Fully Human

THIRD EDITION

CHRISTOPHER UHL
with JENNIFER ANDERSON

FOREWORD BY LAURA E. HAKE

ROWMAN & LITTLEFIELD
Lanham · Boulder · New York · London

Executive Editor: Nancy Roberts
Editorial Assistant: Courtney Packard
Higher Education Channel Manager: Jonathan Raeder

Credits and acknowledgments for material borrowed from other sources, and reproduced with permission, appear on the appropriate pages within the text.

Published by Rowman & Littlefield
An imprint of The Rowman & Littlefield Publishing Group, Inc.
4501 Forbes Boulevard, Suite 200, Lanham, Maryland 20706
www.rowman.com

6 Tinworth Street, London SE11 5AL, United Kingdom

British Library Cataloguing in Publication Information Available

Library of Congress Cataloging-in-Publication Data

Names: Uhl, Christopher, 1949– author.
Title: Developing ecological consciousness : becoming fully human / Christopher Uhl, Penn State University with Jennifer Anderson.
Description: Third edition. | Lanham : Rowman & Littlefield, [2020] | Revised edition of : Developing ecological consciousness : the end of separation / Christopher Uhl. Second edition. [2013] | Includes bibliographical references and index. | Summary: "Rather than working through a list of environmental problems, it aims to help students awake to the awe and wonder of our planet, begin to understand some of the challenges facing it, and explore possibilities for action and change"— Provided by publisher.
Identifiers: LCCN 2019043729 (print) | LCCN 2019043730 (ebook) | ISBN 9781538116685 (cloth) | ISBN 9781538116692 (paperback) | ISBN 9781538116708 (epub)
Subjects: LCSH: Ecology. | Sustainable development. | Restoration ecology. | Ecological assessment (Biology)
Classification: LCC QH541 .U44 2020 (print) | LCC QH541 (ebook) | DDC 577—dc23
LC record available at https://lccn.loc.gov/2019043729
LC ebook record available at https://lccn.loc.gov/2019043730

♾™ The paper used in this publication meets the minimum requirements of American National Standard for Information Sciences—Permanence of Paper for Printed Library Materials, ANSI/NISO Z39.48-1992.

To all of us born into these tumultuous times.
May we turn toward life.

Contents

Foreword

As I gently sit with the invitation to write this foreword, I struggle with how to introduce a book that has forever changed how I relate to myself, other people, and the world. I now live deeply in a way that honors Earth through listening and persistent, gentle curiosity directed toward myself and everyone and everything around me. I learn and feel so much more. I watch spider webs pulse in the wind. I know how to be at peace within my infinitely worthy, interconnected self. I teach from my heart. So, as I begin this writing challenge, I take a few full breaths . . . and a way forward comes to me. By telling you my story about how this book entered and transformed my life, I can invite you to the incredible journey contained herein.

My transformation began quietly in 2013. I'd recently closed down my research program in biomedical sciences at Boston College and was preparing to ramp up my teaching. My faculty life was going okay. Personally, however, I felt anxious and filled with agony over the devastation and degradation happening everywhere: famine, floods, fires, mass extinction. Even at that time, I could see that Earth's ecosystems were unstable and that the world that my son and my students will inherit was in crisis.

In the midst of my heartbreak, I read this:

> The crisis we face is first and foremost a crisis of mind, perception, and values—hence, a challenge to those institutions presuming to shape minds, perceptions, and values. **It is an educational challenge**.[1]

Aha! Here's a way I can move out of passive despondency: I can expand my role as *teacher*. I can choose to develop a course for non-science majors that includes information about dear Earth; that clarifies the reality that we are intimate elements of Earth; and that instills values of compassion, generosity, and curiosity. I was on a mission. I started reading more broadly and talking with colleagues in other departments. I realized that the course needed to be expansive and to spark enthusiasm from a wide range of students. Standard environmental science texts were not nearly broad or deep enough. That's when I found the second edition of *Developing Ecological Consciousness: The End of Separation*. "Developing ecological consciousness"—this phrase promised so much. I understood "developing" as "growing and learning," and "ecological consciousness" implied awareness of, and connectivity to, the living planet to which we are all inextricably bound. As I perused Uhl's table of contents, I found many of the themes I wanted to include in my class: Earth as an intensely interconnected system, ecosystem collapse, and reflection.

I started reading and immediately had the sense that I was in conversation with a big-hearted, dear friend. Part I oriented me to my place in the universe, as a being that arose from Earth. The science was solid (bolstering my confidence in this book as a "science" text) and was embedded in a larger context—enlightened by insights from wisdom teachers, spiritual leaders, and cosmologists. I was struck by how succinctly, naturally, and compassionately the text led me to a realization of my intimate interconnection with Earth and the universe. Then, as I said "yes, yes, yes" to the importance of what I was learning, I discovered the powerful and unique soul of this book: intriguing activities for truly embodying the book's teachings along with provocative questions that challenge readers to dig into the material as well as to reflect upon themselves. *Really interesting!* For example, in chapter 1 I read a compelling and informative account of the history of the universe from vastly different perspectives: the big bang from our astrophysicists and a correlate "let there be light" from the biblical story of Genesis. Might both of these "stories" be explanations of the same first-order mystery? Then I was encouraged to *experience* the universe: I went outside on a very dark night, lay down in the grass, allowing Earth's gravity to keep me from falling into space, and gazed out at a star-studded sky. As prompted, I then brought to mind that hydrogen, the most abundant element in my body, was created in the big bang, at the beginning of time, fourteen billion years ago . . . . Whoa . . . the stuff of me was present at the very beginning.

I then reflected on a question from the end of the chapter: "What does it mean to you to live in a universe?" As I pondered this question, I felt my heart opening and an increased sense of comfort and belonging. I actually began to take deeper, more nourishing breaths.

This book led me to an extremely different way to "learn" about the natural world. My initial reaction to the content, layout, and tone of the book went from "this is weird" to a deep sense that "this is right."

As I moved into part II, "Assessing the Health of Earth," I braced myself for the typical doom and gloom diatribe about how Earth is falling apart. Instead, from the sense of connection I felt leaving part I and the nonjudgmental writing style, I was able to take in the well-supported facts of Earth's breakdown: climate chaos, pollution, degradation of ecosystems. Each scenario was succinctly followed up with the very real, negative impacts that Earth breakdown is having on humans and all living creatures. There was no condescension, no preaching. In fact, the gentle invitation was to learn how to deeply listen to ourselves, to others, and to Earth. Through this listening, I found I could more fully take in what's happening to Earth and all her inhabitants. I realized that when I attend carefully to what's happening around me, I feel more deeply and am able to respond as my authentic self. In addition, I rediscovered the power of curiosity. As my curiosity increased and I opened myself to asking lots of questions about myself, others, and

Earth, my daily anguish and despondency about our imperiled world was actually lessening, and connection to myself was growing.

Given the structure of the book, I was expecting part III to offer up "solutions" to our current crises. What unfolded instead is perhaps the most invigorating part of the text: an invitation to create a new story for our lives, grounded in an understanding of ourselves as intimately interconnected with Earth, of how our culture shapes us, and how, with curiosity and courage, we can explore and, in fact, make choices that honor ourselves, each other, and Earth. By the end of the book, I was jubilant; this text was perfect for my new course!

In late January 2014, I was in my office with tears rolling down my face. I'd started teaching my course in mid-January and was now reading a student reflection on a question from chapter 1, "How does this new understanding of the universe change your understanding of yourself?" She was sharing the tender realization of her connection with life, her shift from feeling inconsequential to feeling intimately connected—and completely worthy—within the cosmos. As I continued to read student reflections over the course of the semester, I discovered more and more vulnerable expressions of love and struggle and humility. After fifteen years of teaching, for the first time, my students became so much more than names and faces and grades. They were now fully dimensional, with fears, worries, and struggles, and so much love. My heart cracked open.

I realized that my powerful private experience with this book was just a beginning: I was on the cusp of a much deeper exploration, in community with my students. These days, during class periods, I lecture a bit, using videos and stories that highlight the text. We then discuss perspectives, consternations, and ideas raised in student reflections on the text and end-of-chapter activities. *Every time I teach this course*, as our conversations broaden and deepen, students experience a turning point when they realize that this class is *very* different. It is about everyone and everything in the cosmos and, at the same time, deeply personal. The book, when they let it, leads them through an exploration of their individual lives and choices as well as circumstances beyond their control. They realize that they have incredible power: to be curious, examine life more deeply, take risks, and make choices grounded in greater understanding. *Yes!* We are in crisis and **YES!** We can do something about it. From that first semester to the present, I continue to be awestruck by the impact this book has on students and, in turn, their impact on me.

I've shared this book with hundreds of students as well as friends and family and have discovered that this book is for ANYONE who wants to take part in the healing of our world, on any scale. Different disciplinary perspectives are interwoven naturally, mutually supporting one another with fact and connection.

This text could easily be used for courses in psychology, theology, philosophy, history, sociology, anthropology, biology, political science, economics, or environmental studies. In fact, although written for college-aged students, the use of this book can be adapted for any group of learners wanting to help our imperiled Earth, from school-aged through adults of all ages, anyone. A community action group, book club, or study circle could use this book for hours of enriching discussion, reflection, and action.

Developing Ecological Consciousness contains the wisdom and gentle guidance needed for opening up to our extraordinary potential as conscious, self-reflective beings. I am boundlessly grateful to have been able to share this path to *becoming fully human* with students, friends, family, and now you.

Welcome to *Developing Ecological Consciousness: Becoming Fully Human.*

Laura E. Hake
Associate Professor, Biology Department, Boston College

Preface to the Third Edition

It is not enough to invent new machines, new regulations, new institutions. We must develop a new understanding of the true purpose of our existence on Earth. Only by making such a fundamental shift will we be able to create new models of behavior and a new set of values for the planet.

—Vaclav Havel[1]

In 1982, I accepted a faculty position in the Biology Department at Penn State University. My time was to be divided between research and teaching. The research part was familiar, but teaching was new. My first semester at Penn State, I was assigned to teach a large environmental science course targeted at non-science majors. Being a first-time teacher, I had never grappled with such basic questions as: What does it mean to teach? How do you do it? How does learning occur? These questions were not on my radar because I, naively, assumed that there was only one way to teach, only one way to learn.

With a standard environmental science text in hand, I plunged in, determined to "cover the material," filling my student's heads with facts and figures. There was certainly no lack of material to cover, especially when it came to threats to the environment. Everywhere I looked, I saw (or read reports of) wounds—for example, forest clear cuts, acid rain, ozone thinning, polluted rivers and oceans, climate warming, toxins in our food, starving children, wars, genocide—a world seemingly hurtling toward its own demise. Reading and lecturing about all of this ecological havoc took a toll on me; I became sad, angry, frustrated, confused.

Can you imagine being me—*Dr. Death*—tracking the deterioration of Earth's vital signs, semester after semester? Or perhaps worse: Can you imagine being a student, sitting in a room with five hundred of your contemporaries, receiving information about how Earth was in the *shitter* and there was probably nothing you could do about it?

At the end of my fifth year, I had a breakthrough that occurred during the last day of the semester while my students sat, heads bent, laboring over their final exams. From my perspective, I'd had my best teaching semester ever, but when my students handed me their exam sheets, only a handful made eye contact and there were no smiles . . . zero thank yous. Yet, only minutes earlier I had been congratulating myself for doing my best job *ever* as a college teacher. Something was clearly amiss.

A few days later, feeling deeply unsettled, I strapped on my backpack and headed into the wilds for a weeklong walkabout. During this wilderness sojourn, I realized that I had been teaching my course upside down because I was asking my students to care about something—Planet Earth—with which most had only limited contact and relationship. After all, for the most part, they lived indoor lives, separated from daily contact with rivers and streams, rocks and ridges, tides and currents, mist and moonlight, mosses and meadows. How absurd of me to expect them to care about a world from which they were separated mentally, physically, emotionally, and spiritually! And how tragic that by grounding my course in negativity and despair, I was cultivating hopelessness, numbness, and further separation.

Out of my despair, a new question emerged: What would happen if I turned my course right-side-up by grounding it in delight, curiosity, awe, connection, self-reflection, and empowerment? Indeed, what if my intention was to help my students fall in love with Earth—to open their senses wide to the wild and wonderful Earth that had birthed them into being?

By the end of that walkabout, I was determined to completely reimagine my course. More questions surfaced: Do I really need to use a standard environmental science textbook? And, if I find the current textbooks arid and soulless, what is keeping me from writing my own book? What if I get rid of exams? How about I abandon the safety of the podium and move around the lecture hall, engaging students close up? What if I build my course around questions—especially students' questions—rather than around answers? In sum, what if I switch my role from information broker to that of midwife—that is, a coach with the mission of helping my students make sense of these tumultuous times so that they might begin to discern their life's meaning and purpose. All of this led me to imagine my course not so much as a subject to be taught as a journey to be taken.

That walkabout, two decades ago, was the catalyst for writing the first edition of *Developing Ecological Consciousness* (DEC), released in 2004. At that time, the concept of *sustainability* was acting as a wakeup call, warning humankind that Earth is finite and that there are ecological limits within which we will need to live if we hope to endure as a species. Because I supported this thesis, I subtitled the first edition of DEC *Path to a Sustainable World*.

Second Edition: The End of Separation

During the nine-year interval between the release of the first edition and the second edition in 2013, the health of Earth's oceans, forests, soils, rivers, and atmosphere continued to deteriorate. Upshot: We were decidedly not on a *Path to a Sustainable World*.[2]

In preparing the second edition, I asked myself, repeatedly: Why are we destroying Earth? What's at the root of our behavior? Is it arrogance? Greed? Stupidity? Blindness? Numbness? It is, I believe, a combination of all of these.

Earth is unraveling before our very eyes because we remain ignorant of the most fundamental reality of human existence—namely, that Earth is primary; that it is her life force that sustains and nurtures us. If each of us carried this awareness in our consciousness, day by day, we couldn't abuse Earth as we now do. Instead, we would extend gratitude and respect to her, from morning 'til night, doing everything in our power to nurture, sustain, and love her—but clearly this is not happening.

In our short-sighted ignorance, we mistakenly assume that Earth belongs to us, but in truth, we are the ones who belong to her. With this truth in mind, I chose the aspirational subtitle, *The End of Separation,* for the second edition of *Developing Ecological Consciousness.*

Third Edition: Becoming Fully Human

The environmental crisis has continued to broaden and deepen since the second edition of DEC was released in 2013. In crafting this new edition, my colleague, Jennifer Anderson, and I have taken pains to scrutinize and update each chapter, each paragraph, each sentence. In addition, we have expanded our treatment of the mind-boggling roles that microbes play in the healthy functioning of our bodies (chapter 2) and the profound impacts of man-made chemicals on human reproductive biology (chapter 5). Along with adding new material in many places, we have removed content that, in our view, lacked poignancy.

Finally, we have organized the content of this new edition around the theme of *Becoming Fully Human.* This subtitle conveys the message that, as a species, we still have some serious growing up to do. Indeed, our continued abuse of and separation from Earth signal that we are still a species very much in our adolescence.

Though we humans pride ourselves on our cognitive intelligence, we will need more than our clever brains to resolve the ecological crisis that we have backed ourselves into. Fortunately, we are more than mere *brains on a stick*: we are inspirited; we are ensouled; we each have the innate ability to feel, to sense, to intuit, to dream, to love, to act with courage. These capacities are essential parts of our human equipment, but as long as we remain cloistered in our heads, holding the fullness of our shared humanity in check, our exploitation and abuse of Earth will continue. Indeed, people from all walks of life are already openly acknowledging that civilization, as we have come to know it, might not last much longer. From where I stand, this assessment is not alarmism but a forthright acknowledgment of the greed, fear, and ignorance that have become emblematic of our times.

If we are to wake up in time to avoid our own demise, we will need to acknowledge that in many ways we are failing to live up to our birthright— that is, failing to discover and manifest what it means to be fully and expansively human; failing to live with deep meaning and purpose; failing to grow up. If you are wondering if this might be true of you, consider this question from poet Mary Oliver: "Are you breathing just a little and calling it a life?"

If your primary mission, as a human being, is to conform to our culture's narrow script for success by devoting your life energy to the pursuit of money, status, and comfort, the chances are good that you are unwittingly "breathing just a little and calling it a life." Why? Because by following this *business-as-usual* life script, you will miss the opportunity to become fully alive, fully human, fully yourself.

We become fully human only to the extent that we learn to use all of our human *equipment*.[3] Take wildness, for example: If you are human, you are an animal, with a particular suite of sensibilities, instincts, and potentialities that are innate to every animal species. Yet, most of us have been so thoroughly socialized—that is, domesticated—that we are reluctant to express the full spectrum of wildness and aliveness that is part of our birthright. How is it for you? How playful are you? How sensuous? How erotic? How instinctive? How spontaneous? Have you ever given yourself permission to hug a tree, to sleep alone under the stars, to weep freely in public, to sing until your heart bursts with joy, to act with fierce courage, to howl at the moon, to forage for wild foods? If not, you are missing opportunities to use all your human equipment . . . missing out on what it means to be fully and unabashedly human.

Spirit is another component of our human equipment that must be cultivated if we are to become fully human. In ecological terms, *spirit* is an energetic sensibility that awakens us to the fact that we belong to—are part of—something that is both extraordinary and deeply mysterious. Modern physics confirms that subtle forces of connectivity and interdependence permeate all of existence, meaning that the human-created concepts of *separation* and *independence* are illusions. Though our culture conditions us to believe that each of us is a skin-encapsulated ego, the larger truth is that everything connects. Indeed, the word *spirit* connotes the *interbeing* of all that is.

Finally, on this journey toward becoming fully human, there is a soul element. In the context of our human equipment, the soul is that ineffable force that whispers to us through dreams and intuition—like an underworld muse—guiding us toward our essence, our destiny, so that we might offer the world the gift, no matter how humble, that is uniquely ours to give. Indeed, it is the persistent itch from our shy souls that calls us toward what most matters in life, beckoning us to express our full humanity and, in so doing, to discover our life's true meaning and purpose so that we might all join together in the healing of the world.

Of course, none of this is easy. In fact, it's a daunting challenge; yet each of us has the necessary equipment. We just need to summon the courage to use it. May this book serve you on your path.

Christopher Uhl
Jennifer Anderson

Acknowledgments

Our thanks go to Sarah Stanton (previously at Rowman & Littlefield), who encouraged us to write this new edition of DEC. Our gratitude also goes to Courtney Packard and Jehanne Schweitzer at Rowman & Littlefield for their skillful means in moving this new edition through the publication process.

We also wish to extend a special thanks to Dr. Laura Hake at Boston College for writing the foreword to this edition and for making significant contributions to chapters 2 and 5.

Finally, our heartfelt gratitude goes to Jean Forsberg for granting us permission to grace this book's cover with one of her extraordinary paintings.

About the Authors

Christopher Uhl: Early in my career, I had an interest in both medicine and ecology, and as my life's work has unfolded, I have been able to join these interests under the banner of "ecological healing." During the 1970s and 1980s, I studied the ways in which Amazon rain forest ecosystems heal after human assaults. Then in the 1990s I focused on the role that universities can play in ecological healing by modeling sustainable practices. More recently I have been engaged in teaching and writing. It was my dismay with prescriptive environmental science texts that prompted me to write the first edition of DEC in 2003. I am also the author, along with Dana Stuchul, of *Teaching as if Life Matters: The Promise of a New Education Culture*, a book that offers a new paradigm for soulful learning and teaching in the tumultuous times in which we find ourselves. Most recently, I have written *Awaken 101: Finding Meaning and Purpose in College and Beyond* (forthcoming), a handbook directed to all those endeavoring to make sense of these frenetic times. See my website at www.chrisuhl.net.

Jennifer Anderson: I am grateful to have grown up near the headlands of the Allegheny River, where the hills seem vast and the communities are tightly knit. I spent much of my childhood roaming, exploring, and playing outdoors, so finding a career path related to these adventures made sense. I am grateful for Penn State University's Shaver's Creek Environmental Center, which has been my professional home for many years. Through several positions there, including the most recent shift to teaching college students, I have been able to do what I love, sharing with others the natural connections that I find so captivating and vital: people seeing and feeling the beautiful interrelationships of Earth, people relating with other people, and people coming to understand themselves more fully. My philosophies are wholeheartedly wrapped up in the content and ideas contained in this book. I am grateful and honored that Chris asked me to help with this third edition of *Developing Ecological Consciousness*, as it brings so many of my own ideas and sensitivities into print.

PART I
Earth, Our Home

The real voyage of discovery consists not in seeking new landscapes but in having new eyes.

—Marcel Proust[1]

The goal of part I is to connect—specifically, to forge connections with the cosmos, Planet Earth, and the life around us. By connecting in these ways we open ourselves to potentially life-altering experiences such as those described by some of the early astronauts. For example, when astronaut Rusty Schweickart was released from his space capsule on an *umbilical cord* during an early Apollo mission, he looked back to Earth, "a shining green gem against a totally black backdrop" and realized that all that he loved was on that gem: his family; the land and rivers of his home place; art, history, culture. This jet-fighter pilot was so overcome with emotion that he wanted to "hug and kiss that gem like a mother does her firstborn child."[2]

Schweickart had another breakthrough: As he observed Earth from space, he saw that clouds did not stop at national borders to check for political ideology because the natural world is built on the principles of connectivity and interdependence, not separation and independence.

The call to recognize our interdependence by forging direct connections to the natural world is unnerving to many people, young and old. You, yourself, might be thinking: "I already am connected! I know what I am doing!" If this is true for you, consider these questions: What makes the world go around? What keeps you alive? Is it money? Technology? Business? Government? God? Really, what is it that supports your life?

In pondering this question, did you happen to consider the four most fundamental elements of your support system:

1. sun—that great radiance that lights up the world;
2. water—the miracle liquid that is the medium for life;
3. soil—the living matrix that is the substrate for life; and
4. atmosphere—that all-permeating elixir that we draw our breath from.

If you failed this test, you are not alone. As we have become more specialized and dependent on technology, our separation from Earth has increased. This separation is literal, insofar as we spend more and more time indoors, living inside *boxes*. Indeed, we start each day inside a box (house or apartment). Next, we hustle off to work or school in small horizontally moving boxes (cars, buses, trains). Then, we often spend our days in office boxes, school boxes, factory boxes, business boxes. In the evening, we play it backward, moving from box to box, returning home. Amazingly, it's possible to pass an entire day without any direct contact with the living Earth: never feeling unfiltered sunlight on our skin or cool breezes ruffling our hair; never hearing the sounds of free-running water or walking on unpaved surfaces; never plucking a ripe berry from a bush or delighting in the flight-dance of starlings at dusk.[3]

We pay a price for this insofar as living our lives in boxes disconnects us from Earth. No surprise, then, that most of us have become indoor people with indoor concerns, connected to our televisions, phones, and computers, our wild selves increasingly ransomed to a simulacra existence.

This need not be. Part I of this book is an invitation to see Earth with new eyes—the eyes of relationship—and, in so doing, to assume our birthright as vital and awakened members of the Earth community.

1

Humility

We Are a Part of Something Greater Than Ourselves

> Indigenous peoples . . . live in a universe, in a cosmological order, whereas we, the people of the industrial world, no longer live in a universe. We in North America live in a political world, a nation, a business world, an economic order, a cultural tradition, a Disney dreamland.
>
> —Thomas Berry[1]

When I first read Thomas Berry's words, I was confused. How could it be that "we, the people of the industrial world, no longer live in a universe?" But if you and I have no day-to-day awareness—that is, no embodied experience—of living within a universe, then how can we say that we do? And to the extent that we are blind to the fact that we live in a universe, is there anything we can do to change this? This question lies at the core of this opening chapter.

Some years ago, having spent far too much time in the *industrial world*—the *business world*—I retreated to the forested mountains of northern Pennsylvania, where I established a camp, tended a fire, and roamed free in the woodlands by day. At night I slept in a hammock slung between two pines and listened to owls hoot and coyotes yelp before falling into fitful slumber.

I recall one evening in particular, when I was awakened by the need to pee. Rising from my hammock, I stumbled into the darkness to relieve myself. When I finished, I was disoriented and stumbled about—hands stretched out in front—reaching for my hammock strings, but to no avail. Feeling anxious, I took a deep breath and considered that the worst that could happen was that I would spend a few hours, on a star-strewn night, free to contemplate the heavens.

I looked up and saw the Big Dipper showing clearly through an opening in the forest canopy. Then I held two pine needles in a crossed position so that their intersection demarcated a single point within the Big Dipper's bowl. With this gesture, I was mimicking what the Hubble space telescope had done when astronomers had focused its lens on a speck of sky equivalent to the space that was defined by my two crossed pine needles, all the while taking in the faint light from distant galaxies. The result was an extraordinary image that registered light from some two thousand distinct galaxies.

To survey the entire bowl of the Big Dipper, Hubble would need to take roughly twenty-five thousand photographs! An examination of this intensity would reveal an estimated forty million galaxies in the patch of sky demarcated by the Dipper's bowl; an assessment of the entire sky, in this manner, would turn up more than one hundred billion galaxies—each galaxy with billions of stars, many of those with families of planets. In this context, Boston College professor Chet Raymo calls on us to

> [g]o outside and hold those crossed [needles] against the night sky. Let your imagination drift away from the Earth into those yawning depths where galaxies whirl like snowflakes in a storm. [Then] from somewhere out there among the myriad galaxies, imagine looking back to the one dancing flake that is . . . our Milky Way.[2]

Following Raymo's call, I was, for a brief time that evening, aware that I was *living in a universe.*

Today, by joining discoveries in the sciences together with humanity's great wisdom traditions, we are able, as never before, to fathom the workings of the Earth and Sun that sustain us as well as the origins of the complexifying and expanding universe that is our larger home. As we awaken to this magnificence—to the stupefying mystery of it all—how can we not be humbled?

Foundation 1.1: The Origins and Workings of the Universe

To be human is to be curious. At an early age, we begin to ask questions, for example, "How did I come to be? Why is the sky blue? Where did Planet Earth come from?" Though it is our human nature to ask questions, I confess, with a mixture of regret and embarrassment, that I taught environmental science for ten years before it occurred to me to invite my students to ponder how our topic of study, the environment—that is, the world beyond the boundary of our skins—came into existence.

As a way of gaining some perspective on this question, put yourself back thirty thousand years to the time when our hunter-gatherer ancestors inhabited Earth. If you had lived then, what story would you have created to explain all those points of light (stars) in the night sky? Perhaps those flickering specks were the campfires of your ancestors—souls who had died and passed into the spirit world? And how far away were those "campfires?" A day's walk? A year's walk? How could you figure it out?

The first Greek astronomers, puzzling over this, observed the Sun circling the Earth each day; at night, they saw the moon and stars move across the sky. Based on their Earth-centric perspective, they created the celestial spheres hypothesis, which posited that Earth was nested at the very center of a series of concentric spheres. The Moon, they thought, was embedded in one sphere, the Sun in another, Mars in another, and so forth, all the way out to the final sphere that contained all the stars of the night sky.

Those early astronomers assumed that the multitude of stars embedded in the star sphere differed in their relative brightness because they were of different sizes—bright stars large, dim ones small. The possibility that the dimness of stars might owe to their remoteness in space was not seriously considered. Absent any notion of deep space, it seems reasonable that the early Greeks imagined that all the stars embedded in the star sphere were only a short distance out from Earth.[3] Living within this mental construct, the only object of significant size in the universe would have been Earth, which of course would explain why they believed that Earth stood at the center, with everything else circling around.

But then along came Copernicus, Galileo, Kepler, and others. Based on their studies of the heavens, using telescopes and mathematics, these pioneering scientists blew the lid off the celestial spheres hypothesis, positing that it was the Sun, not the Earth, that was at the center of the universe and that the Sun was millions of miles away. For those of us alive today, this all seems rather obvious, but put yourself back five hundred years. One evening

> [y]ou go out under the stars to consider these ideas. You know the stars well for they are your calendar and nighttime clock. You stand at the center of your familiar universe, looking up at the well-known, nearby stars, when suddenly—it can happen no other way—the celestial sphere shatters and the stars are hurled at varying, unknown distances.[4]

In that moment you grasp, for the first time, that a star's brightness or dimness is related to how far away it is, and you stand awestruck as you consider that there might be stars so far away that they are too faint to see. Your world is no longer small; the universe may be too big to see, and humans are no longer at the center. Feel the dissonance and discomfort as your old worldview collapses, but also, feel the excitement as you are beckoned into a new, mind-bogglingly expansive worldview.

A New Story

Just as people in the 1600s resisted the new knowledge presented to them by science, so it is that today we sometimes find it difficult to embrace scientific discoveries. For example, we readily accept that Earth revolves around the

---

ONE HUNDRED BILLION GALAXIES!

Imagine that you were assigned the task of naming all of the galaxies in the known universe, and suppose that you were able to name a new galaxy every second, day and night, without sleeping. Even with these advantages you would still need more than three thousand years to give names to all the galaxies in the known universe!

Sun (even though from our Earth-bound perspective, the opposite appears to be the case), but most of us struggle to comprehend that our Sun is just one of tens of billions of stars in the Milky Way galaxy and that this galaxy is just one of the hundred-billion-plus galaxies that make up the known universe.

It has been humbling for humans to acknowledge that Earth is not at the center of the universe, and what's more, that Earth is only a smallish planet circling a smallish star (our Sun) that exists way out toward the edge of a galaxy (the Milky Way) that happens to be just one among more than a hundred billion distinct galaxies in the known universe. And yet, in a fascinating way, our galaxy—all galaxies, really—is/are at the center because the universe is *omnicentric*, meaning that no matter your location in the universe, you are, in a sense, at the center.

To understand this paradox, consider first that all the known galaxies (and galaxy clusters) in the universe are moving away from each other. One way to visualize this is to think of a loaf of raisin bread baking in an oven. Let each raisin represent a galaxy. Now place yourself on one of those raisins. As the bread bakes and rises and as you look around, you will see all the other raisins (galaxies) moving away from you. You would perceive the same thing if you were positioned on any of the other raisins. Thus, no matter which raisin (galaxy) you were on, you would experience yourself at the center.[5]

Not only are galaxies expanding apart from each other, but the farther they are from each other, the faster they go. Therefore, galaxies that are twice as far apart are separating two times faster, and those that are ten times farther apart are racing away ten times faster. Space, literally, rushes into existence as galaxies separate from one another.

The Big Bang
Employing the laws of physics, modern-day scientists have been able to trace this process of expansion backward to a hypothesized beginning point—the birth of the universe!—estimated to have occurred roughly fourteen billion years ago. Think of it as watching a movie that begins in the present and then runs back through time, as the billions upon billions of galaxies in today's universe reverse their outward expansion and begin to draw together, even-

A FOURTEEN-BILLION-YEAR-OLD UNIVERSE!

The notion of fourteen billion years, just like the idea of one hundred billion galaxies, is hard to fathom. See if this helps: Imagine that one millimeter (roughly the width of the dot on this "i") is equivalent to one year. Thus, one meter (composed of one thousand millimeters) would equal one thousand years, and one kilometer would correspond to a million years. Using this scale, the entire fourteen-billion-year history of the universe would span a distance from Washington, DC, to Tokyo. Now, imagine yourself standing at the base of the Washington Monument. Your first step (about a half meter) would take you back five hundred years to the time when the Spanish first sailed to America. Four more steps and you would be back to the time of ancient Rome. Trekking a few dozen miles into northern Virginia, you would be back in the age of the dinosaurs (65 million years ago). To arrive at the time when the Earth and Sun were formed (roughly 4.6 billion years ago), you would have to walk all the way to the California coast. And if you were truly earnest about traveling to the beginning of time (back to the big bang), you would still have to make your way across the immense expanse of the Pacific Ocean, all the way to Tokyo.[6]

tually merging into a hot, high-density gas. With further contraction and astoundingly high temperatures, all the atomic mass is converted to pure energy as the entire universe collapses into the infinitely small, infinitely hot, *singularity* that gave rise to the big bang fourteen billion years ago.[7]

Scientists don't simply "run the movie backward" and then accept that the universe began from a big bang fourteen billion years ago. They search for corroborating evidence. For example, here on Earth, they do experiments using high-energy particle accelerators to understand what happens to matter at excruciatingly high temperatures. Then, using this information, they can calculate the nature of the early universe that emerged from the big bang. These calculations predict that the matter that formed from the big bang would have consisted of hydrogen and helium in a ratio of three to one, which is, in fact, the material makeup of the universe today as determined by astrophysicists.[8] Furthermore, if there really was a big bang, it stands to reason that the faint light from that ancient cosmic explosion should still be detectable. Although this radiation wouldn't be visible to the naked eye, scientists reasoned that it could be measured with infrared and radio telescopes. In fact, in 1989, the satellite dubbed COBE (Cosmic Background Explorer) measured this radiation with great precision and found that its temperature and behavior matched that predicted by big bang theory. Here's the punch line: The big bang isn't lost far back in time; today Planet Earth receives light that was emitted when the universe was first forming. The big bang is all around us and in us![9]

The Big Bang in Us?

I remember learning in high school that my body was 70 percent water and that without my skeleton to hold me upright, I'd be a watery bag sloshing around on the ground. Another fact I absorbed—not very interesting at the time—was that a water molecule is composed of two atoms of hydrogen plus one atom of oxygen. Only much later, when I began to wonder about the actual origins of hydrogen and oxygen, did I learn something truly astonishing: All the hydrogen now on Earth, including all the hydrogen in each of our bodies, was created fourteen billion years ago when the universe burst into being. Hydrogen was created *only* in that big bang moment—never again since—*and* it was hydrogen that subsequently gave rise to the first galaxies and first stars.

Consider the implications of this the next time you hold a glass of water. Given that all the hydrogen in that water is fourteen billion years old and that hydrogen is the most abundant atom in your body, how old are you? Your human age might be fifteen or fifty, but the essence of you, the stuff of you, is fourteen billion years old. In an atomic sense, it appears that we've all been here since the very beginning.

But perhaps you are wondering: What caused the big bang in the first place? When physicists consider this question, they explain that it is theoretically possible for something to materialize out of a vacuum—out of nothingness! According to MIT physicist Alan Guth, a speck of matter one-billionth the size of a proton may have bubbled into being and created a repulsive gravitational field—a false vacuum—so strong that it exploded into our universe.

The phenomena known as "black holes" may offer a clue to all of this. For example, some scientists speculate that black holes—rather than being places of death and demise—might actually act as birthing sites for new universes. For example, astrophysicist John Gribbin speculates that the big bang may have been a moment in time/space when our universe was birthed out

---

A MULTIVERSE?[10]

While she was a graduate student at MIT, Katie Bouman led a team of researchers in the development of the mathematical equations that helped capture the first image of a black hole. The image was released in 2019. The black hole in this now famous photo is five hundred million trillion kilometers away from Earth and measures forty billion kilometers across. That's three million times the size of the Earth! The image was captured by a network of eight telescopes that scanned a galaxy named Messier 87 for ten days. Professor Heino Falcke of Radboud University in the Netherlands observed: "The black hole is an absolute monster, the heavyweight champion of black holes in the Universe."

of a preexisting black hole in *another* universe. Viewed through this lens, it could be that our universe is just one among many universes, all linked together by "tunnels," akin to cosmic umbilical cords. The idea of a family of universes, or a "multiverse," though difficult to prove, is considered a distinct possibility by cosmologists today.[11]

Foundation 1.2: Origins and Workings of the Solar System

The famous astronomer and cosmologist Carl Sagan observed that we are all composed of stardust. By learning how stars, such as our Sun, and planets, such as Earth, are formed we can now appreciate the literal truth in Sagan's words.

Formation of Stars and Planets

Our Sun came into being roughly 4.6 billion years ago, and you may be surprised to learn that it owes its existence to the *death* of preexisting stars that were formed when the universe was still relatively young. As those *first-generation stars* burned up all the hydrogen fuel in their cores, they generated heavier elements such as carbon, nitrogen, oxygen, calcium, and silicon that were released in a matrix of dust and gas. In cases where dying stars were really large, their death was marked by enormous explosions called *supernovae* that created the conditions for the birth of second-generation solar systems. Upshot: Without star death, new stars, like our Sun, and planets, like Earth, would not exist.

Use your imagination now to go back five billion years and picture the gas and dust, generated by a supernova explosion, gathering and collecting in gigantic whirlpools and eddies with hydrogen atoms clumping together at the center. Sense the enormous attractions and pressures that eventually led the hydrogen atoms to fuse into helium atoms, provoking massive outbursts of energy that caused the center to ignite, forming a star! First, there was a cloud of hydrogen atoms; then, because of the immensity of the forces that came into play, our Sun was born.

And what about the formation of Earth? How did that happen? Here is how Brian Swimme and Mary Evelyn Tucker describe it in their lovely book *Journey of the Universe*:

> In the beginning, our infant sun was completely surrounded by hydrogen, carbon, silicon and other elements disbursed by supernova explosions. As they drifted through space these elements would brush against each other and begin to cohere into tiny balls of dust. Over millions of years, these "planetesimals" continued accreting to each other and growing until they

were the size of boulders and then as large as mountains. Not all collisions resulted in larger bodies. Many were so violent that they tore both bodies apart. But over millions of years these planetesimals continued to absorb all the loose material circling about. [So it was that] our solar system, with its eight planets, its band of asteroids, and its one infant sun, slowly came into being.

It is remarkable to realize that, over immense spans of time, stellar dust became planet. In the earliest time of the universe, this stellar dust did not even exist because the elements had not yet been formed by the stars. Yet hidden in this cosmic dust was the immense potentiality for bringing forth mountains and rivers, oyster shells and blue butterflies.[12]

As for you and me—every part of our bodies contains this stellar dust; we are stardust through and through.

The Wondrous Sun

It is easy to take the Sun for granted; it has been burning brightly for almost five billion years, and astrophysicists calculate that it still has five billion years of hydrogen fuel left to burn. Only in the twentieth century did scientists come to understand how the Sun works. It is an extraordinary story that offers a window into Einstein's equation, $E = mc^2$, which describes how mass is converted to energy. According to this equation, the amount of energy (E) produced by the Sun is related to solar mass (m) multiplied by the speed of light, squared (c^2). What a remarkably simple equation—two things multiplied together to give an answer! And one of those two things, the speed of light (c), is already known—186,000 miles per second. The only thing missing to solve the equation is solar mass (m). Yes, our Sun provides Earth energy by continuously sacrificing its own mass. This occurs at the Sun's center, where astounding temperatures cause protons to fuse together. Two protons are reluctant to fuse because their positive charges repel each other ferociously. Here on Earth, for example, this union is only possible in enormously expensive particle accelerators or in the fury of atomic bomb explosions.[13] But the immense size of stars, like our sun, creates the massive gravitational forces and associated temperatures that allow for nuclear fusion. Now here is the key part: As protons fuse at the Sun's core, they lose a small amount—about 1 percent—of their mass. This lost mass is the *m* in $E = mc^2$. Though the amount of lost mass in these fusion events is tiny, when it is multiplied by the speed of light squared (i.e., by approximately 35 billion), the result is large.

The energy that is released from proton fusion at the Sun's core percolates upward until it reaches the Sun's roiling surface, where it is hurled into space as heat-light. Eight billion pounds of the Sun's proton mass is lost each second. That's right, eight billion pounds! Of this vast quantity, Earth intercepts just five pounds' worth. Think of it as five pounds of transformed solar mass arriving on Earth as radiant energy each second, lighting up and powering Earth. Five pounds of solar mass doesn't sound like much until we appreciate the wonder of $E = mc^2$. In fact, the amount of solar energy intercepted by Earth in just one hour is greater than the total amount of fossil fuel energy burned by humankind in one year.[14]

Without the Sun's warming radiation, temperatures on Earth would be hundreds of degrees below zero—far too cold for life to exist; without the Sun's rays, there would be no evaporation of water from oceans and leaf surfaces, and that would mean no rainfall, no hydrological cycle! Nor would there be photosynthesis. Yes, the Sun is, in effect, a giant generator that powers the winds, the water cycle, photosynthesis, and ultimately, our very bodies. No wonder so many ancient peoples regarded the Sun with a sense of awe and reverence!

Recalling the entirety of this epic story, we see that our solar system—coalescing into existence from cosmic gas and dust generated by star explosions—has now given birth to us. Incredibly, our very bodies, our cells, and the air we breathe are the work of the heavens—gifts from the stars!

The Origins of Life on Earth

By remarkable good fortune, owing to its distance from the Sun and its size, Earth offers conditions that are ideal for life. For example, Earth is not so close to the Sun that its surface is scalded, as is the case with Mercury, nor is Earth so far from the Sun that its life-giving water is perpetually frozen, as is the case with Jupiter. Likewise, Earth is not so large that its gravity holds a dense blanket of atmospheric gases that filter out sunlight, nor is it so small that it fails to hold onto a life-sustaining atmosphere.[15]

Although Earth is well situated at present, it came into existence as a fiery magma ball. As that ball slowly cooled over millions of years, vapor condensed into clouds and rain fell and fell for so long that oceans formed.[16]

Likely, it was in Earth's primordial seas that life first appeared, but nobody knows how. Was it simply by chance? Was there some as yet undiscovered self-organizing principle involved? One science-based theory of life's origins, known as *panspermia*, holds that life didn't arise on Earth at all, but somewhere else, out in space. According to this view, the primitive "seeds" of life—think bacterial cells—are adrift throughout our galaxy; by chance, they occasionally colonize a planet like ours, arriving via meteorites or comets.

Another theory proposes that the first complex organic molecules on Earth were produced with energy contributed by lightning and/or by the release of heat from deep-sea vents. To test the plausibility of this theory, scientists placed water—to represent the early oceans—in closed glass containers and added the gases that were presumably present in Earth's early atmosphere. Then they applied heat and electric sparks to simulate lightning and, in so doing, were able to generate some common chemicals found in plants and animals, for example, amino acids, fatty acids, and urea; but of course a collection of organic molecules, by itself, doesn't constitute life.[17]

Somehow, the molecules for life—whether originating here on Earth or in space—became encased in a cell. How? It is possible, as alluded to previously, that among the many substances in Earth's early ocean *soup* were oils, such as fatty acids and lipids, that produce bubbles (cells) when mixed with sea water. Based on this premise, microbiologist Lynn Margulis speculates:

> In the earliest days of the still lifeless Earth, such bubble enclosures separated inside from outside. Pre-life, with a suitable source of energy inside a greasy membrane, grew chemically complex. . . . These lipid bags grew and developed self-maintenance . . . [and over time] acquired the ability to replicate more or less accurately.[18]

Suffice it to say that at this point in time, the origins of life remain shrouded in mystery.

The Diversification of Life on Earth

Earth's first organisms, appearing approximately four billion years ago, were primitive forms of bacteria. It has been only in the last 10 percent of Earth's history—starting some six hundred million years ago—that more complex life forms have appeared, such as mosses, ferns, fungi, crustaceans, insects, amphibians, reptiles, and flowering plants. The first mammals came into existence just two hundred million years ago.

Although humans are latecomers in this evolutionary saga, our origins actually extend all the way back to the appearance of single-celled, bacterial life. That's right: Our earliest ancestors were not primitive primates; they were bacteria! This might sound crazy, but how could it be otherwise? Life arose, evolved, and complexified until eventually our hominid form arose, but it all began with bacteria. And let's not forget the stars, because we wouldn't be here were it not for those stars that expired long ago, creating, in their dying, the materials that gave rise to our solar system and with it, those first bacteria that have led to *us*!

Life on Other Planets?

Given the abundance of life on Earth and the unfathomable size of the universe, what is the probability that life might exist on other planets? Though we don't know for sure, it seems unlikely that we are alone in the universe. Take the case of the Milky Way, one of an estimated one hundred billion galaxies that make up the known universe. Our Sun is one among one hundred billion–plus stars in the Milky Way. That's one hundred billion other suns, each with the potential to have its own solar system; but to be conservative, let's imagine that only half those suns actually have planets. What remains is roughly fifty billion suns, each with planets, in the Milky Way galaxy.

However, planets with the characteristics that make them candidates for life—for example, being just the right distance from their sun and just the right size for the formation of a life-sustaining atmosphere—might be relatively rare. So again, let's be conservative and grant that only one in every five hundred of those fifty billion suns in our galaxy has a planet with characteristics required for life as we know it. With this math, we still have roughly a hundred million planets with the potential for life support in just *our* galaxy.

With the possibility of extraterrestrial life seemingly all around us, perhaps a more appropriate question is: Why haven't we had any visitors? Or if we are having visitations, why aren't they more common? The reason, in large part, may be related to the vast size of our galaxy. For example, the average distance between stars, considering all stars, is five hundred light years. And how far is that? One light year is the distance that light, moving at the speed of 186,000 miles per second, travels in one year. And how far is that? Let's do the math: 186,000 miles/second × 60 seconds/minute × 60 minutes/hour × 24 hours/day × 365 days/year = 5,865,696,000,000 miles/year. Yes, one light year is approximately six trillion miles—a colossal distance—and the average distance between stars is not one light year, but five hundred light years! And, of course, the average distance between that tiny subset of stars with potentially life-supporting planets would be even greater. In addition to these vast distances in space, we also have "distance" in time—that is, civilizations comprised of intelligent life would be scattered in time. To grasp this concept, consider that our solar system has been around for almost five billion years, but intelligent life (like us), capable of reaching out into space, has been in existence for only a minuscule fraction of that time. Thus, at any one time we may have only a few hundred stars in our galaxy simultaneously supporting a planet with life, and the average distance between these stars, given the size of the Milky Way, would likely be on the order of thousands of light years.[19] Upshot: It is likely that there are other forms of sentient life in the universe, but visitations appear unlikely.

Stepping Back to See the Big Picture

My body and the universe come from the same source, obey the same rhythms, flash with the same storms of electromagnetic activity. . . . So, it must be that the universe is living and breathing through me. I am an expression of everything in existence.

—Deepak Chopra[20]

Our ancestors, back through time, created stories to make sense of the world. Today this process continues as cosmologists weave discoveries regarding the origins and evolution of the universe into stories. A starting point for today's cosmology is the observation that our universe is structured by gravity—a force of attraction that keeps objects connected to one another. Gravity seems to have been a factor from the very beginning of time. For example, current mathematical models of the early universe assert that elementary particles (i.e., quarks, leptons, Higgs boson) came into being at the moment of the big bang. These primordial particles quickly morphed into protons and neutrons and then commenced to bond together in stable relationships. Yes, from the very first moments, it appears that our universe has been creating relationships:

> Things could have been different. We can theorize about a different kind of universe, a universe that would have taken the form of disconnected particles, a universe that would never have formed bonded relationships. . . . But in our observable universe, various forms of bonding are inescapable. . . . Bonding is at the heart of matter.[21]

I find it reassuring to live in a universe grounded in relationship. Likewise, when I learn that our Sun sends five pounds of itself to Earth each second—so that we can all live—I feel a mix of wonderment and gratitude, recognizing that a forerunner of human generosity is stellar generosity.[22] And when I grasp that the most common element of my body, hydrogen, originated fourteen billion years ago at the birth of our universe, I stand a little taller knowing that my roots are ancient. Finally, when I learn that stars die in stupendous explosions (supernovae) that promote creative outbursts—that is, the creation of new solar systems—I realize that I live in a profoundly dynamic universe.

For me, the most exciting thing about the universe story is that it is not over. We live in the midst of an unfurling universe, and as such, we are participants in an epic story. This story tells us that hydrogen was created in the big bang and that now, fourteen billion years later, hydrogen has differentiated and complexified into corn plants, butterflies, lizards, and people! In effect, energetic and gravitational forces have acted on hydrogen atoms over billions of years, transforming this element into all that is.

This story invites us to recognize that the universe is not a thing, nor is it a place; rather, it is a dazzling process of emergence with an inborn tendency to self-organize, generating more complexity, more consciousness, and more aliveness over time.

But What about God?

If you believe in God(s), the revelations from this modern science-based story of creation need not undermine or even contradict your faith. Indeed, revelations from science could just as easily serve to deepen and broaden your spirituality. For example, rather than understanding God as a *being*—a noun—you could choose to understand God as more akin to a verb—an action or process—for example, God as the animating force that infuses all beings: human being, animal being, plant being, atmospheric being, geologic being. This new way of seeing sets aside the old idea that God created the world for us, challenging us to consider, instead, that humans have been made for the world. In other words, it is *not the world that belongs to us*, but *we who belong to the world*. This shift in worldview presents a profoundly humbling challenge to our human identity—one that promotes connection rather than separation, peace rather than war, respect rather than abuse, generosity rather than greed.[23]

In sum, science and religion, rather than being antagonistic, have the potential to become mutually enriching. Indeed, you could use the scientific information presented in this chapter to make inferences about the nature of God (or your placeholder for God). How? By considering questions like the following:

- What kind of creator would place coupling and relationship at the center the universe?
- What kind of creator would make a universe that is wondrous and mysterious beyond our wildest imaginings—that is, a universe that to our eyes is both unimaginably large *and* imperceptibly small?[24]
- What kind of creator would fashion a universe where destruction (think supernovae) leads to rebirth and transformation?
- What kind of creator would manifest a universe that is not fixed or static but dynamic—in a constant process of moving toward greater complexity, greater aliveness, and greater intimacy?

As I playfully explore these questions, words like *loving, compassionate, generative, spontaneous, whimsical, wise,* and *benevolent* arise as descriptors of the ultimately unnameable force(s) underlying creation.[25]

Wrap-Up: Expanding Ecological Consciousness

It's all a question of story. We are in trouble just now because we do not have a good story. We are in between stories. The old story, the account of how the world came to be and how we fit into it, is no longer effective. Yet, we have not learned [a] new story. . . . We need a story that will educate us, a story that will heal, guide, and discipline us.

—Thomas Berry[26]

For better or worse, stories shape our beliefs, our worldview, and ultimately, our well-being. Some people alive today live in a story that implies that we live in a universe that is indifferent, remote, dead, cold, and lacking in meaning and vitality. Caught in this story, it is possible to conclude that in the big picture, our existence as humans is pretty much a pointless accident and our best strategy, while we are alive, is to seek comfort any way we can. The upshot of this *dead universe* story, as educator Duane Elgin points out, is "a tendency toward materialism, hedonism and the exploitation of nature."[27]

However, it now appears, based on recent work in physics, astronomy, and cosmology, that the universe is *anything but dead*. Instead, it is alive with energy. Beneath the solid surface of material objects, there is a mind-blowing flow of energetic activity. This energy can be appreciated at the scale of a single atom, as the electrons that circle the nucleus of each atom are vibrating several trillion times a second.[28] Yet in spite of all this movement, the universe presents itself as a unified whole. For example, scientific experiments reveal that subatomic particles (e.g., photons) "respond" instantaneously to each other, irrespective of the distance that separates them—that is, if one particle is changed, the other particle will register a corresponding response instantaneously, even though that "other" particle might be a million miles away. Einstein referred to this phenomenon as "spooky action at a distance"; today scientists call it *quantum entanglement*. Physicists have demonstrated that it is real, though they remain in the dark as to how it works, but that it occurs at all suggests that the universe exists as a single undivided whole.[29] Upshot: Rather than a dead universe, it seems that we live in a participatory universe where everything connects.

In the end, of course, each of us gets to choose our story. Choosing to see ourselves as dwelling in a universe that is lifeless and unresponsive readily engenders feelings of indifference, separation, arrogance, loneliness, and fear. But choosing to see ourselves as a part of a universe that is fecund, alive, intelligent, and profoundly creative evokes feelings of connection, curiosity, awe, respect, humility, and love. The choice is ours.

Applications and Practices: Cultivating Humility

> Just as the Milky Way is the universe in the form of a galaxy and an orchid is the universe in the form of a flower, we are the universe in the form of a human. And every time we are drawn to look up into the night sky and reflect on the awesome beauty of the universe, we are actually the universe reflecting on itself. And this changes everything.
>
> —Brian Swimme[30]

Discovering Our Place in the Universe

Here is a culminating question for you to ponder: Where is the universe? Point to it. Seriously, where is it? When given this prompt, most people point away from themselves, out into space. It seldom occurs to us to point inward because we are not accustomed to seeing ourselves as participants in the universe's unfolding—that is, we have little awareness that we are an embodiment of the universe. But all the calcium in your bones, the iron in your blood, and the carbon in your tissues had its origins in the bellies of stars—stars that in their dying seeded the heavens with the elements necessary for life. Both your body and the universe are composed of the same elements; they come from the same source.

Spending most of our time indoors, as so many of us now do, we easily forget that the universe is our larger home. We can *re-member* ourselves home by engaging in the age-old practice of stargazing. On a cloud-free night, pack a blanket and a thermos and head out alone or with a friend to a quiet spot in the country, away from the glare of artificial lights. Then, spread out your blanket, lie down, and gaze up into the night sky. In so doing, you will be looking at the Milky Way galaxy from within its boundaries.

If you could somehow remove yourself from the Milky Way and look down on it from outside, it would appear like a gigantic Frisbee—one hundred thousand light years across—with a bulge at the center (see figure 1.1).[31] But the Milky Way is not out there, separate from you; you live within its great wheel of stars.

As you lie there taking in your home galaxy, you may feel stationary, but actually, you are traveling in several dimensions. As you already know, Earth

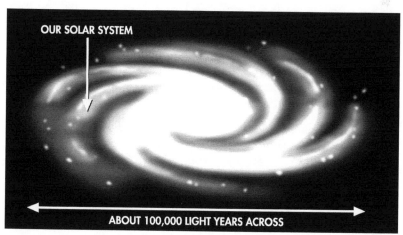

OUR SOLAR SYSTEM

ABOUT 100,000 LIGHT YEARS ACROSS

Figure 1.1 A representation of the Milky Way galaxy.

is spinning on her axis each day and orbiting around the Sun each year, but did you know that our solar system is also involved in a grand galactic rotation whereby it spirals around the vast Milky Way galaxy once every 225 million years. And don't forget that the Milky Way (with you included) is still racing outward from the big bang!

One more thing to contemplate: As you lie on your back, it is natural to assume that you are looking *up* at the stars, but cosmologist Brian Swimme reminds us that "up" is just a cultural construct. Neither Earth nor the Milky Way has an *up* or a *down*. When you stand on Earth's surface, you are not standing up; rather, you are sticking out into space. So, as you lie on your back, instead of thinking of yourself as looking up, picture it so that you are on the underside of Earth looking down into the blackness of the night sky. It may take a while, but eventually you will experience all the stars as way down there below you, and you may even marvel at the fact that you are not tumbling down into the space below you. You don't fall, of course, because Earth's gravitational pull holds you. Not your weight but the Earth's hold suspends you above the stars. If Earth's gravitational embrace were to suddenly vanish, you would tumble into that dark chasm of stars below you:

> As you lie there feeling yourself hovering within this gravitational bond while peering down at the billions of stars drifting in the infinite chasm of space, you will have entered an experience of the universe that is not just human and not just biological. You will have entered a relationship from a galactic perspective, becoming for a moment a part of the Milky Way Galaxy experiencing what it is like to be the Milky Way Galaxy.[32]

Sky gazing can be like seeing a movie—you go to it once and say, "Been there, done that"—but it can also become a practice. Devoted sky watchers know that no two nights are the same. Atmospheric conditions, time of night, phase of the moon, time of year, and physical location all affect what can be seen. Binoculars can enhance the experience. But why stop there? The next time you go stargazing, consider listening to Joseph Haydn's *Creation Oratorio* on your phone. Get settled by lying down on a blanket, closing your eyes, and breathing yourself into a relaxed state . . . and then press play:

> Silence. A C-minor chord, somber, out of nowhere. Followed by fragments of music. Clarinet. Oboe. A trumpet note. A stroke of timpani. A prelude of shadowy notes and thrusting chords, by which Haydn meant to represent the darkness and chaos that preceded the creation of the world. . . . Listen now, eyes closed, as the music descends into hushed silence. . . . Open your eyes! A brilliant fortissimo C-major chord! A sunburst of sound. Radiant. Dispelling darkness. A universe blazes into existence, arching from horizon to horizon. Stars. Planets. The luminous river of the Milky Way. As you open your eyes to Haydn's fortissimo chord and

to the (almost) forgotten glory of a truly dark starry night, you will feel that you have been a witness to the big bang.[33]

Sadly, at present many of us are much more likely to be sitting in front of a television (or some other screen) at night rather than beholding a star-strewn sky. Screens seduce us to remain indoors, dwelling in a world of human artifice. The wild night sky, by contrast, calls forth both awe and humility, reminding us that we are a part of something that is both precious and mysterious. Walk alone, with humility, into the dark night; behold the heavens; go quiet; you are a part of something much greater than yourself.

Questions for Reflection

Throughout this book, at the end of each chapter we offer six questions for reflection.

- What impact does the universe story (as summarized in this chapter) have on your understanding of who you are and on your life's meaning and purpose?
- How does seeing and experiencing the universe—not as a static entity, but as a dynamic happening—leave you feeling?
- How does the story of Earth's origins, as related in this chapter, affect your understanding of concepts like *spirit, god, mystery*?
- What might you do to acknowledge your utter dependence on the Sun's benevolence?
- In your experience, how might behaving with humility expand your awareness? Your consciousness?
- Finally: What does it mean for *you* to live in a universe?

These questions are invitations to *reflect* on chapter 1 within the context of your own life. We don't call them "reaction questions," because *reactions* are conditioned responses that arise, absent serious thought. A reflection question, on the other hand, requires that we take time to ponder, trusting that the question, when taken seriously, will expand our consciousness. Here is a four-step warm-up exercise for exploring the difference between *reacting* and *reflecting*.

Step 1. Pick a question from the list that intrigues you. For illustrative purposes, let's say you chose the final question "What does it mean for *you* to live in a universe?" Pause right now and respond to this question by writing down the first thoughts/reactions that come up. Give this two minutes.

Step 2. Reflect on your Step 1 reactions by asking yourself: Why did I react to the question (What would it mean for me to live in a universe?) the way I did? Then, go further and consider what your reactions reveal to you about yourself. Responding to these twin questions will require more effort than Step 1. See this as an exercise in self-study, an opportunity to be intimate with yourself. Give it five minutes.

Step 3. When you are ready, return to the question, "What does it mean for *you* to live in a universe?," but this time, rather than simply reacting as you did in Step 1, engage in genuine reflection by exploring questions like: (1) What role does the universe play in my life? (2) How might I initiate a relationship with the universe? (3) What might a conversation between me and the universe sound like?

Give this a full ten minutes, firm in the conviction that you have within you a deeper, more profound response than what you came up with in Step 1. Write down what you discover. Then take a break, allowing the question to marinate in your psyche for a day.

Step 4. Take a walk out into the night, and when the time seems right, lie down and gaze up into the cavernous night sky. As you take in the heavens, ask yourself: "What *does* it mean to me to live in a universe?" Live this question by paying attention to the thoughts and feelings that arise. As you settle in more deeply, bring your attention to your breath, the breath that is bequeathed to you by the very universe that you are beholding.

Conclude this exercise by reviewing your experiences as you went from *reacting* to the question (Step 1) to *reflecting* on it (Steps 2 and 3), and finally, to exploring it from a place of embodied receptivity (Step 4).

Be gentle with yourself. Reflection is challenging, but the self-discovery that it promises makes it worth the effort.

2

Curiosity and Connection
Seeing the World with New Eyes

> In reality, there is a single integral community of the Earth that
> includes all its component members whether human or other than
> human. In this community, every being has its own role to fulfill,
> its own dignity, its inner spontaneity. . . . Every being enters into
> communion with other beings. This capacity for relatedness, for
> presence to other beings, for spontaneity in action, is a capacity
> possessed by every mode of being throughout the entire universe.
>
> —Thomas Berry[1]

Driving along a desolate stretch of road in the Everglades in 1972,
I encountered a young couple hitchhiking. Seeing their bedrag-
gled appearance, I picked them up and asked where they had
spent the night.

The guy, Dave, responded, "We crashed back in the 'glades. Ya know,
trying to get back to nature."

"Whoa!" I exclaimed, "Must-a-been a lot of mosquitoes? How much
'Off' did you use?"

Dave's friend, Sally, said softly, "We don't use repellents."

I probed a bit more, suggesting, "You must have had your blood
sucked dry?"

Dave said with a sigh, "Yeah, we got bitten up pretty good but, ya'
know, it's all related."

Then, Sally fixed me in her gaze and added, "When you really think
about it, everything is connected—it's all connected."

I sensed that those two were saying something important, but I
couldn't quite grasp it at the time. In retrospect, I see that they were challenging me to see nature, not as something oppositional and separate from
me, but instead as something that we are all a part of—something to which
we all belong.

How would you characterize your relationship with nature? For example, what happens when you leave your fabricated "indoor" world behind
and step into the *wall-less*, wild world that is nature? Like Sally, do you feel

at home, like you belong? Or does being *out* in nature—for example, walking alone in a forest—leave you feeling a bit jumpy and uncomfortable? Or might it be that you simply experience the outside world as a backdrop for your life—like pleasant wallpaper?

There is no judgment intended with these questions. Insofar as most of us spend most of our lives indoors, it is no wonder that we might think of nature as something "out there," apart from us. But what if it could be otherwise? This is where Sally's words, "It's all connected" and this book's subtitle (*Becoming Fully Human*) come together.

Foundation 2.1: Nature All Around Us

Early in my time as a graduate student at Michigan State, I took a field ecology course with Dr. Patricia Werner. On the first day of class, Dr. Werner led us out into a meadow and told us to spend fifteen minutes just looking around at what was there. That's all she said.

Place yourself there in that meadow, with fifteen minutes to "just look around." What would you do? Where would you look? Would you walk around surveying nature from above, or would you settle down onto the ground, engulfed by nature's touch, nature's sounds, nature's smells?

When our time was up, Dr. Werner gathered us together and, gesturing with her arms to encompass the entire meadow, asked: "What's going on here?" That was it. No elaboration. Just: "What's going on here?" I assumed that there was a specific answer that she was fishing for, but I couldn't fathom what that answer might be. So I remained silent. In retrospect, I now realize that Werner was introducing us to the *scientific method*.

Though you probably don't see yourself as a scientist, you are engaged in the four-step *scientific method* any time that you make an *observation* (Step 1) that triggers a *question* (Step 2) that results in a provisional answer or *hypothesis* (Step 3) that you can test through *experimentation* (Step 4). For example, suppose you want to enter a building. You see (*observation*) that there is a door and you wonder (*question*) if you should open the door by pulling it or by pushing it (*two hypotheses*). You decide to push (*experiment 1*), and the door remains stationary. Then you pull (*experiment 2*), and the door opens. In quick succession, you have enacted the four-step scientific method, going from observation to question to hypothesis to test. In this sense, we all have opportunities to act as scientists in our everyday lives. We begin when we take time to simply observe our surroundings, as Werner was inviting me to do in the meadow that day.

If you remain dubious about my assumption that we all possess the cognitive equipment necessary for science, you can test it by spending ten minutes with a tree, simply observing (Step 1 of the scientific method). Do this, and your innate curiosity will soon have you asking questions based on your observations. Start with the tree's bark, examining it closely, noting colors, textures, patterns, variations. These observations will trigger questions that, very likely, will begin with words like *why, how, when,* and *what.* Then, playfully consider possible answers (*hypotheses*) to one of your questions. Finally, add in creativity by imagining a way that you could *test* one of your hypotheses. If you feel intimidated by this challenge, just remind yourself that, as humans, we all possess the essential dispositions necessary for science: namely curiosity, imagination, playfulness, and creativity.

Seeing a Butterfly with New Eyes

As a way of activating your *inner scientist,* picture yourself on a summer afternoon walking through a meadow and encountering an orange-red butterfly perched on a wildflower. Bedazzled, you wonder, "What's going on here?" Looking closely, you note that the butterfly's wings are scribed with dark lines and that the wing margins are black with white spots. After a bit of Googling, you determine that this butterfly is a monarch.[2]

Now, dear reader, in the spirit of play, use your imagination to shrink yourself down to the size of a sand grain—that is, small enough to fit on the monarch's head, between her two antennae. I grant you that this suggestion may sound a bit ridiculous, but when scientists give free rein to their imagination, changes in perspective often lead to insights.

From that position, perched on the monarch's head, you note that she stops frequently to lay eggs on the leaves of milkweed plants. This observation prompts you to ask, "How does the monarch distinguish a milkweed from the scores of other plant species in the meadow?" Does she use sight . . . smell . . . touch . . . something else?

With a bit more Internet sleuthing, you learn that monarchs use scent receptors on their antennae to smell their way in the general direction of milkweed plants. Then, once they land on a candidate plant, they employ taste receptors located on their *feet* to verify that it's a milkweed.

Curious to know what milkweed tastes like, you rip off a piece of leaf, but when you see milky fluid emerging from the tear, you decide against a taste test. Later, you learn that biochemists have determined that the milky leaf fluid contains cardiac glycosides—toxins that cause heart paralysis and severe vomiting responses in vertebrates. You and I would become ill if we nibbled at a milkweed leaf, and yet monarch caterpillars can eat milkweed leaves absent negative effects, because they have the capacity to isolate and excrete milkweed's toxins.

But wait! Why do monarchs go to all the trouble of ingesting a plant loaded with toxins when there are scads of other, more palatable plants to choose from? In pursuing an answer to this question, you learn that not only are monarch caterpillars able to excrete milkweed's toxic glycosides,

CHANGING PERSPECTIVE

As a novice scientist conducting research in the Amazon basin in the early 1970s, I was taken to a huge swath of land where giant bulldozers had scraped away the forest, leaving no trace of plant life. Two weeks later I returned to that desolate moonscape, curious to know if any plants had been able to colonize. I did find tiny seedlings, here and there, and I marked each one with a small tag; but the following week, when I checked on them, they had all died.

When I shared this information with an older, more experienced, scientist, she challenged me to imagine what it would be like to be a tiny seedling attempting to survive in a barren landscape. Intrigued by her suggestion, I pictured myself as a tiny seedling, feeling the hot sun sucking the moisture out of my tiny leaves. Then, longing for relief, I visualized a heavy rain falling. But with my new seedling eyes, I saw the raindrops that came crashing down, creating mini-craters in the soil, exposing my fragile roots to the burning sun. It was then that it occurred to me that in addition to nurturing seedlings, rain, when heavy, could actually kill them. With further observations and tests, I confirmed this hypothesis. Though mine was a rather elementary discovery, I would not have stumbled upon it had I not endeavored to imagine how plant seedlings experience their world.

Later, I learned that German biologist Jakob von Uexküll invented the word *umwelt* to describe the distinct ways that each species experiences the world.[3] For example, when you and I walk through a patch of forest, we experience our surroundings in ways that are defined and limited by our human sense organs. Meanwhile, other forest species (e.g., deer, vole, raven, earthworm, oak, rhododendron) have their own unique sensory systems, leading them to experience the forest in ways that are distinct from us. Even plants (though they lack a nervous system) respond to light, water, airborne chemicals, and physical touch in ways that are unique to them and beyond our human capacities.[4] Acknowledging that each life form has its particular way of sensing the world—that is, its own *umwelt*—is a wonderful way to cultivate imagination, awe, and empathy.

they can also store these toxins in their tissues. In fact, biologists have discovered that the concentration of glycosides is higher in monarch tissue than in the actual milkweed leaves that these butterflies eat. Because monarch caterpillars carry this toxin and are brightly colored, birds and other animals that prey on caterpillars quickly learn to avoid monarchs. Indeed, researchers have demonstrated that when blue jays are offered a variety of caterpillar types, they quickly learn to reject the monarch caterpillars.

There is more to this story! Fast forward: The life cycle of the female monarch that you have been accompanying is almost over. As she lays her last egg, you slide down from your spot between her antennae and nestle in beside that last egg, curious to learn what will happen to it. After four days, a tiny caterpillar hatches and begins eating the milkweed plant. As she feeds and grows, she undergoes five molts over a three-week period; then, fat and full, she affixes herself to a branch and forms a mummy-like container (i.e., pupa) within which her body is deconstructed, turning to "soup," and then reconstructed into an adult monarch butterfly.

When the butterfly emerges from her chamber, you are curious to see what's next, and so again, you find a perch between her antennae, holding on tight as she spirals high into the sky in a courtship flight. You are still with her when she settles down in the late afternoon with a male suitor, and you watch as this suitor inserts a packet of sperm, complete with nutrient supplements, into a special receptacle in her reproductive tract. The next day you accompany her as she begins to lay eggs, releasing sperm to fertilize each egg just before it is laid.

The entire monarch life cycle is repeated three times each summer, with new generations emerging, reproducing, and dying every six weeks. But what happens in the winter? Specifically, how do the monarchs that hatch in fall survive through the winter? Many insect species stay put and survive as dormant eggs or encased pupae, but monarchs have a different strategy; they migrate south.

For many years the migratory destination of monarch butterflies was a mystery, until Dr. and Mrs. F. A. Urquhart decided to mark a bunch of monarchs born in the fall in Michigan. They did this by affixing tiny adhesive tags to the wings of hundreds and hundreds of monarchs. Each tag contained a minuscule number and a mailing address, indicating where to send the monarch if s/he was found. So, for example, if one of the monarchs tagged by the Urquharts in Michigan was hit by a car in Arkansas and the driver discovered this tagged monarch on the car's grill and sent it back, the Urquharts would know that at least one of the Michigan monarchs that they had tagged went as far south as Arkansas. Using this tagging technique, the Urquharts, along with other researchers, pieced together the migratory route of the monarchs, eventually determining that they overwinter, by the millions, in a mountain reserve just west of Mexico City (see figure 2.1).

Flying approximately fifty miles per day, the monarchs survive their journey to Mexico by feeding on flower nectar along the way. In fact, researchers have discovered that monarchs actually gain weight during their trip south by converting flower nectar into fat, which they store in their abdomen. These accumulated fat reserves ensure the monarch's survival during their three-month stay in the Mexican highlands, where little food is available to them. At their winter roosts, the monarchs cluster by the thousands on the branches of fir trees. The dense fir woodlands protect them from freezing while keeping them cool enough to remain semidormant.

At the end of March, they rouse themselves and slowly return north, feeding as they go, until they make it to the southern United States, where

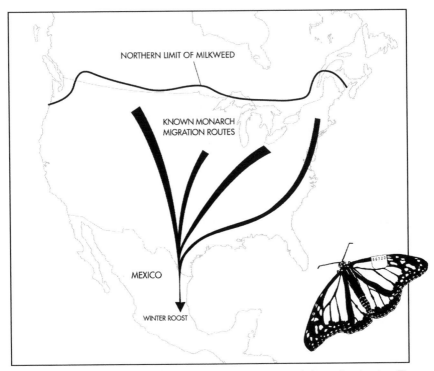

Figure 2.1 Monarch butterfly migration routes. Note monarch butterfly wing tag. These tags help scientists determine the monarch's migratory movements.

they mate and lay eggs. Then, exhausted by their epic journey, the adults die, leaving it to their progeny to complete the trip to their summer breeding grounds in the northern United States, where three generations of monarchs emerge, reproduce, and die, leaving it to the fall-born monarchs to find a way back to Mexico for the winter.

If, after reading this, you are curious about monarchs, you can nurture this interest by joining thousands of others at Monarch Watch (http:// monarchwatch.org), where amateur naturalists, acting as citizen scientists, report on monarch migratory movements. For example, monarch watchers in Mexico might send out the word in early spring that the monarchs are beginning to leave their mountain roosts. Then, observers along the U.S. border might begin to post messages such as "monarchs just arrived here," "monarchs laying eggs," "monarch caterpillars less abundant than usual," and so forth. You, too, could be one of these watchful citizens sending messages to other monarch lovers.

The more we become curious about other species, the more questions we ask and the more we stand to learn. A question currently baffling monarch watchers, as well as scientists in general, is: How do the monarchs that hatch in the fall in the United States find their way back to their tiny wintering grounds near Mexico City, especially given that the fall-hatching monarchs have never been to Mexico?[5] What do you think? Can you come up with one hypothesis?

Seeing Beyond Categories

Our human tendency to categorize what we see by giving everything a name may, paradoxically, interfere with our ability to directly experience nature. For example, suppose one day you pass a sunflower growing in a field and say to your friend, "Hey, look, a sunflower!" Though you have called attention to the sunflower, it is quite possible that you didn't *really* see it. Instead, your cataloged memory of *sunflower* was triggered. But suppose you invited your friend to experience *that* particular sunflower plant, as if neither of you had ever truly been with one before? Your exploration might begin by taking in the whole of the flower head. In so doing, you would surely notice that the head is made up of a multitude of tiny individual flowerets, some closed, some open, some ripe with seeds. Next, with attention on your sense of smell, you might pick up the subtle fragrances, aware that volatile molecules released from the blossoms were merging with your olfactory cells, allowing you to literally take the sunflower's essence into your body. Then, closing your eyes to experience the sunflower through touch, you might discover that sunflower stems are covered in short bristles, as are the backs of sunflower leaves. This observation might lead you to wonder if these bristles have a specific purpose—for example, to discourage insects from feeding on sunflower stems and leaves.

A Field Trip

Using our senses to take in the natural world can be a first step toward relationship. With this in mind, each fall I take my students to a wild meadow ablaze with goldenrod plants. Upon arrival, I ceremoniously give each student a hand lens (i.e., magnifying glass) with the instruction to choose a goldenrod plant that they would like to befriend. Then I challenge them to explore that goldenrod from top to bottom, with the goal of forging a relationship with *her*. Of course, it seems more natural to say, "with the goal of forging a relationship with *it*." After all, we humans typically refer to other life forms as *its*—for example, I cut *it* down, referring to a tree; *it's crawling*, referring to a worm; *it's* blooming referring to a goldenrod. But, by referring to other species as *its*, we objectify them, failing to respect them as unique and vital members of Earth's family of life.

My final coaching to students is to study the goldenrod plant that they have chosen, not as a solitary, stand-alone object, but as an ecosystem, inhabited by pollinators, herbivores, predators, and more.

So imagine yourself in that field, standing before a goldenrod plant. All you see, initially, is one species—the lone goldenrod. Your attention is drawn first to her blossoms; after ten minutes of intense searching, to your amazement, you have had face-to-face encounters with other beings living among her flowers—including various types of mites, aphids, bees, butterflies, beetles, and flies—each feeding, in their own manner, on nectar, pollen, and sap (see figure 2.2). As you become more skilled in the use of your hand lens, you also discover tiny predators, such as ambush bugs and crab spiders, camouflaged among the blossoms, waiting for an opportunity to pounce on an unsuspecting pollinator.

You complete your exploration with a quick survey of the goldenrod's stem, but there's not much happening there, compared to the activity in the blossoms . . . until you suddenly come upon a cluster of aphids. At first the aphids appear to be dead, but with the aid of your hand lens you see that these tiny insects are using their needle-like proboscises to *sip* sap from the goldenrod's stem. As you look more closely, you are shocked to come upon several ants that appear to be slurping up drops of fluid leaking from the abdomens of the aphids. Plant chemists, curious about this phenomenon, have determined that the sap that aphids extract from goldenrods is a good source of carbohydrates but is lacking in the nitrogen that aphids require for protein synthesis. Hence, to satisfy their nitrogen needs, aphids must suck up more sap than they need, releasing the surplus sugar solution from their anuses. The extra sugar doesn't go to waste, as ants, in effect, *milk* aphids. This is a good thing for both parties insofar as it provides ants with an easy source of carbs. Meanwhile, the ants protect the aphids from attack by ladybugs and other potential predators.

Just when you think there is nothing more to discover, you come upon an odd marble-sized globe of plant tissue, called *goldenrod gall,* at the base of the goldenrod stem (see figure 2.2). Curious about what's inside, you slice the gall in half and, in so doing, expose the larva of an insect that you later discover is a *gallfly.* After another Google search, you learn that had you visited this meadow in the spring, you would have seen adult goldenrod gallflies emerging from galls just like the one you have cut open. These gallflies would have been buzzing around the meadow, laying eggs on the tips of young goldenrod plants. The larvae from those eggs would have tunneled down into goldenrod stems and at some point tricked the goldenrod plants (through some fancy biochemical trickery) to create the protective capsule (gall) that serves as both a home and a food source for gallflies during the summer.

In the fall, the larvae in the galls enter winter dormancy; come spring they undergo metamorphosis and emerge as adult flies to mate and repeat the cycle.

Pretty sweet life for the gallfly larvae! Or is it? Even when housed within their snug gall homes, things can get dicey. For example, the lives of gallfly

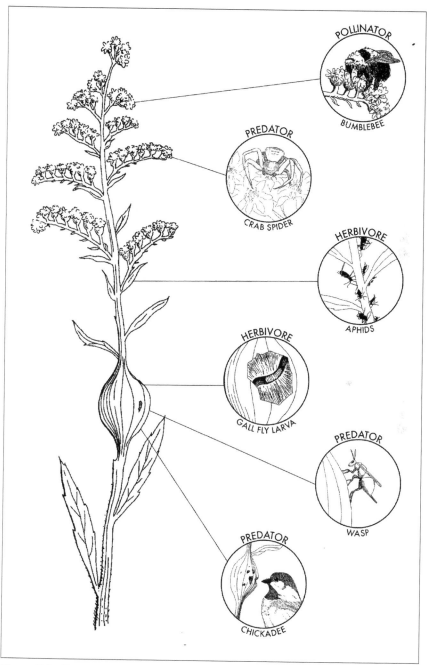

Figure 2.2 Pollinators, herbivores, and predators on a goldenrod plant.

larvae are sometimes cut short in summer by a species of wasp that deposits its eggs inside the galls. The wasp larvae that hatch feed on gallfly larvae. Even if gallfly larvae are fortunate enough to escape attack by wasp predators, they might still be gobbled up in winter by a hungry chickadee or woodpecker, pecking into the gall in search of a meal.

As evidenced in this account, goldenrod plants aren't solitary individuals so much as communities, inhabited by pollinators (bees, butterflies, and beetles), predators (spiders, wasps, ambush bugs, ants, and birds), herbivores (caterpillars and aphids), and more.[6] Any patch of nature will reveal these hidden layers provided we exercise our curiosity.

Foundation 2.2: Nature Within Us

Nature's creatures are not just all around us; they live on us and in us. For example, it is likely that, while you are reading this, there are tiny, scale-covered, wormlike mites feeding in the follicles at the base of your eyelashes. These mites use their needle-like mouthparts to feed on secretions of salt, water, oil, and dead skin from around our eyelashes. At maturity, they mate, and the females drop their eggs into our eyelash follicles. The whole cycle, from egg to adult, takes about two weeks.[7] Some people are grossed out when they learn about their eyelash mites, but here's another way of seeing it: These mites clean our eyelashes by removing dirt and oils and, in return for this service, we provide them with a place to live. The arrangement between humans and eyelash mites is sometimes cited as an example of symbiosis.

Symbiosis is especially common between humans and microbes. In the last half century, microbiologists have discovered that we each have more than a thousand distinct species of microbes living on us and in us. Yes, though you may think of yourself as a separate individual, belonging to the species *Homo sapiens*, your body is a swirling, pulsating ecosystem filled with a wide assortment of creatures. Most of these micro-animals are bacteria, so tiny that a thousand fit on the head of a pin.

Bacteria, as a group, have a remarkable repertoire of capabilities; some can "swim like animals, photosynthesize like plants and decompose dead organic matter like fungi."[8] Indeed, it is bacteria that have pioneered into existence all of life's most essential metabolic processes, for example, photosynthesis, oxygen metabolism, fermentation, and nitrogen fixation.

Insofar as bacteria have been maintaining a vast global network—exchanging genetic material, adapting, and innovating—for almost four billion years, the category "bacteria" is a legitimate one-word answer to the question "What is life?" Indeed, as Lynn Margulis and Dorion Sagan remind us, "Any organism, if not itself a bacterium, is then a descendent—one way or another—of a bacterium. . . . Bacteria initially populated the planet and have never relinquished their hold."[9] The clear implication is that our human evolutionary ancestry traces back to Earth's first bacteria and that "we humans live on a planet that is run for and by invisible microbes. . . . They created our biosphere. . . . They made the soil. . . . They set the conditions for the evolution of multicellular life . . . including us."[10]

Right now bacteria, by the millions, are all around us. They are on our phones, in our water bottles, on our hands before we wash them and after we wash them. No surprise, then, that bacteria inhabit the outside surfaces of our bodies, with hundreds of thousands per square inch on our legs, arms, trunks, and faces.

This abundance pales in comparison with the microbes living inside our bodies. Just in your mouth, alone, there are a variety of distinct species of bacteria. Some of these species are adapted to the environment at the tip of your tongue, some prefer the back of your tongue, others inhabit your cheek pouches, and so forth.

But even this accounting only scratches the surface. In fact, if you were able to count all the cells that comprise your human body (estimated at thirty trillion) and then tally up all of the microbes living in your body, the microbes would actually outnumber your human cells. So, though you may think of yourself as a separate individual, you are, more accurately, a walking bacterial colony. And when it comes to DNA, the master molecule directing your body processes, a mere 1 percent of the DNA contained within and on your body is human; the other 99 percent resides within the bacteria, viruses, and fungi that reside in and on your body.[11]

When we look at a computer, we see the computer's screen and keyboard (the visible body), often failing to acknowledge the computer's essence, its operating system. Similarly, when we look at ourselves, we often fail to recognize and appreciate our invisible bacterial *operating system*. Might it be that our fear of bacteria, to the extent that we experience this, is really our fear of accepting our kinship and dependence on bacteria?

How it is with you? Are you on friendly terms with your bacteria housemates, or do you tend to associate them with dirt and disease? If the latter, that's not surprising, insofar as most of us have been socialized to view bacteria as our enemies.

But rather than regarding bacteria as villains (the vast majority are not!), what if we were to expand our consciousness to view them for what they overwhelmingly are: remarkable beings, essential for our health and well-being?

Our human partnership with microbes is both ancient and essential. We provide a home (i.e., habitat) for them and receive, in return, a multitude of benefits. For example, the massive colony of microbes residing in your gut—the "gut microbiome"—maintains your well-being by (1) breaking down complex carbohydrates into simple sugars that your body can absorb,

(2) synthesizing vitamin K, (3) manufacturing essential amino acids (the building blocks for proteins), (4) detoxifying foreign chemicals, (5) tuning your immune systems, and (6) keeping harmful microbes in check.[12]

The label *gut brain* has been coined to characterize the massive neural communication system that receives and processes information from our abdominal organs and digestive tract and relays this information to our cranial brain. Your gut brain, composed of more than five hundred million neurons that branch throughout your abdomen, weighs three pounds (about the same weight as your cranial brain).

Our gut microbes have a direct line of influence on our cranial brain insofar as our gut and cranial brains communicate with each other. The major conduit for all this communication is the massively complex vagus nerve that extends down from our cranial brain to our heart and lungs before ramifying into the organs that comprise our gut (e.g., liver, kidneys, pancreas, intestines). Though information flows both ways along the vagus nerve gut-brain axis, there are roughly six times more "messages" going from the gut to the cranium than vice versa.[13]

This elaborate information exchange network is essential because digestion is an extraordinarily complex process involving the coordination of a huge range of nutrient needs from within every part of the human body.

Your Gut Microbiome and Your Health

The potential influences of our gut microbiome on our overall health begin at the moment of birth. For example, just as no two humans share identical fingerprints, each of us possesses a unique gut microbiota. As developing fetuses in the womb, we all started out sterile—that is, microbe free. But in the birthing process, microbes selected by evolution to be in our mother's birth canal colonized our bodies. Those of us not born vaginally received a different microbial flora at birth (drawn mostly from our mother's skin). You might think this makes no difference, but evidence is mounting that suggests that C-section babies are more prone to health problems, including allergies and asthma, than are children born vaginally.[14]

Other studies provide evidence that the composition of our gut microbiota could influence our levels of contentment. For example, almost all our serotonin and much of our dopamine—two neurotransmitters associated with feelings of pleasure and well-being—are produced by microbes that reside in the human gut. Current research shows that by regulating the production and release of these two neurotransmitters, our gut bacteria may exercise the power to impact our mood (e.g., whether we are anxious or relaxed), as well as our behaviors (whether we are withdrawn or outgoing).[15] For example, when researchers at McMaster University inoculated bold mice with gut microbes from shy mice, the bold mice became timid and shy. And when they performed the reciprocal experiment by inoculating the shy mice with gut bacteria from the bold mice, the shy mice became more gregarious and adventuresome.[16] Similarly, when the intestinal microbes of obese mice are put into the sterile (microbe-free) bodies of lean mice, the lean mice become obese.[17]

You may be thinking, "These are just mice!" Yet, they are mammals, just like you and me, and their biology reflects our own. One final example: When the guts of normal-size mice that are free of bacteria are populated with intestinal microbes from obese humans, the mice accumulate body fat, whereas *control* mice that receive gut microbes from thin humans show no change.[18]

The implications of this research are staggering. "Just as the Copernican Revolution in the 16th century fundamentally changed our understanding of the world's position in the solar system and Darwin's revolutionary theory of evolution proposed in the 19th century has forever changed our position within the animal kingdom, [today's] human micro-biome science is forcing us again to reevaluate our position on earth."[19] Indeed, how can the fact that the bacterial cells in our bodies outnumber our human cells not have an impact on our consciousness?

Though some may regard these new discoveries as an attack on their identity, we could choose to see them as further proof that there is no such thing as separation on Planet Earth; everything is truly connected and interdependent, even though many of the connections remain invisible to us. Seen in this more expansive way, our gut becomes a precious garden where we can cocreate, with our microbe partners, a healthy ecosystem.

As we learn to tend this garden, it is vital to understand that the microbes in our gut tend to feed on distinct food types. Some microbe species tend to specialize on vegetables, some on fibers, some on fats; others home in on starches, and so forth. This diversity means that if your diet consists mostly of highly processed, high carb, fatty foods, you will create conditions in your gut that favor microbe species that crave these generally unhealthy foods. Indeed, some scientists have found evidence suggesting that microbes might be sending signals to our brains influencing our food choices—for example, creating cravings for food rich in sugars and fats.[20] All of this matters because diets consisting of highly processed foods result in the less diverse gut microbial communities that are now implicated in a wide range of human ailments, including allergies, diabetes, obesity, autism, and depression.[21]

The wisest response for those seeking health and wholeness is to adopt a diverse diet, consisting of a broad mix of vegetables, fruits, grains, nuts, and fibers, while limiting, to the degree possible, the consumption of processed foods and meats. Adhering to such a wholesome diet ensures a diverse, healthy gut flora, and with this comes a vibrant, healthy body and balanced mood.[22] Upshot: Our well-being is dependent, to a significant degree, on our ability to expand our consciousness, coming to see the microbes living in us as precious and essential life partners.

Stepping Back to See a Bigger Picture

> I think the most important question facing humanity is, "Is the universe a friendly place?"
>
> —Albert Einstein[23]

Just as microbes may be influencing our diet, moods, and behaviors, the humans around us may also affect us in surprising ways. For example, a person sequestered in a nearby room could subtly affect your body temperature and your level of calm (versus anxiety) without your being aware of this. This possibility has already been scientifically documented numerous times in an elegant experimental setup involving one person acting as a "Receiver" and a second person acting as a "Sender." Sender and Receiver are isolated from each other in separate soundproof rooms so that they cannot communicate by any known sense—that is, interacting via hearing, sight, taste, smell, and touch are not possible. The Sender's job is simply to focus mental attention on the Receiver during randomly selected thirty-second intervals. When attention is not being focused on the Receiver, the Sender simply relaxes. Meanwhile, the Receiver is instructed to maintain his/her mind in an open state, avoiding any mental fixations. Before beginning, electrodes are placed on the Receiver's fingertips to monitor changes in electrical resistance of the skin, because this measurement is correlated with emotional arousal.

The first studies of this type were conducted in the late 1970s at the Mind Science Foundation in San Antonio, Texas, and the results were highly significant, meaning that Senders did, indeed, influence the emotional arousal of Receivers at a rate much higher than would be expected by chance alone. Since that time this same study has been repeated nineteen times in laboratories throughout the United States and Europe, and the overall results, again, have been highly significant.[24]

Scientists, though they should know better, sometimes discount strange findings, like those described here, because there is no rational explanation for them within our current worldview, but many phenomena that we accept as commonplace today would have been deemed impossible in the recent past. For example:

> Television and cell phones would have seemed miraculous to an eighteenth-century scientist, knowing nothing about electromagnetic fields. Hearing the voices of people far away would have seemed like the work of witches or the delusions of lunatics. But, now, this is an everyday experience, thanks to radios and phones. Likewise, hydrogen bombs would have been unthinkable for nineteenth-century scientists. In the age of steam and gunpowder, such devices would have sounded like apocalyptic fantasies. Lasers would have sounded like mythic swords of light. Indeed, they only

became conceivable for twentieth-century physicists through the scientific revolutions wrought by the theory of relativity and quantum theory.[25]

There's more: There is now evidence that we humans may even be able to influence the functioning of machines. The research comes from the work of Robert Jahn and Brenda Dunne at Princeton's Engineering Anomalies Research (PEAR) laboratory. In one carefully controlled series of experiments, these two scientists used a simple machine that generates a random set of zeroes and ones—like heads and tails. In a series of a thousand numbers, the machine randomly generates approximately five hundred ones and five hundred zeroes, and the probability that the totals might differ from 50:50 (e.g., 490 ones vs. 510 zeroes) can be calculated with precision. When left to run unattended, the machine prints out a completely random list of zeroes and ones, just as one would expect. However, when the researchers invited human beings to sit in front of the machine and to use their thoughts and intentions to bias the machine's output toward more zeroes or more ones, they found that in many cases the machine's output was no longer random. These Princeton "experiments have been reproduced by sixty-eight different investigators in a total of 597 studies. When the results of all these studies are included, the odds against chance being the explanation for [the anomalous results] are 1 in 10^{35} (i.e., 1 followed by 35 zeros)."[26]

An interesting side note to this research is that there was considerable variation among different individuals in their capacity to influence the machines. Those individuals who were most successful described their experience in terms of entering into resonance by merging or bonding with the machine. The very highest scores were achieved by couples who shared a deep love for each other and who worked in tandem to influence the machine. Their scores were up to eight times higher than those of individuals who worked alone.

In interpreting these results, Dr. Larry Dossey writes:

> Proposing that things [e.g., computers] can behave well or badly, or that they have any behavior at all, sounds preposterous to anyone schooled in twentieth-century science. Yet *things* appear to be avid accumulators of significances, particularly when they are chronically exposed to humans. These *things* appear capable of *catching* our feelings, resonating with our emotions, responding to our meanings, and obtruding into our lives, often when we least expect it."[27]

These new findings from science challenge the notion that there is such a thing as a separate, autonomous entity, suggesting, instead, that everything is entangled—human and nonhuman alike—in ways that we are only just beginning to understand.

Wrap-Up: Expanding Ecological Consciousness

> How can we be so poor as to define ourselves as an ego tied in a sack of skin? . . . We are the relationships we share; we are that process of relating; we are, whether we like it or not, permeable—physically, emotionally, spiritually, experientially—to our surroundings. I am the bluebirds and nuthatches that nest here each spring, and they, too, are me. Not metaphorically, but in all physical truth. I am no more than the bond between us. I am only so beautiful as the character of my relationships, only so rich as I enrich those around me, only so alive as I enliven those I greet.
>
> —Derrick Jensen[28]

Life's essence is relationship. Earth throbs with relationship. Yes, that monarch butterfly laying her eggs on that milkweed plant, as well as those microbes digesting fiber in your gut—all creatures large and small—are part of a *worldwide web* of relationship. Science is now revealing all of this to us. Nonetheless, there is still a strong human tendency to regard our own species as separate from, and superior to, all other life forms on Earth. This belief has been coined "speciesism."

What do you think? Just because we happen to have large cranial brains and the ability to communicate complex messages and make tools, does this mean that we are *superior* to the millions of other species that inhabit Earth alongside us? All life forms, after all, have their unique capacities. Plants have the ability to snag solar photons, speeding by at 186,000 miles per second, and then to use those photons to transmute water and carbon dioxide into food. I've never met a human who could do that! Or how about the lowly horseshoe crab, a species that has managed to survive the rigors of Earth's changing environments for more than a hundred million years? If longevity were the primary criterion for superiority, humans—who have been around for a fraction of a million years—would be markedly inferior to horseshoe crabs.

Insofar as assessments of superiority or inferiority are based on arbitrary criteria, they are pretty much meaningless. We can sidestep this bias by acknowledging that humans are not superior or inferior to the rest of the family of life. Rather, the process of evolution has produced Earth's *tree of life*. This tree has a multitude of branches, each its own evolutionary line. The buds along each of these branches are species. Our species, *Homo sapiens*, represents one bud on one branch—no better, no worse, than the millions of other buds distributed along this tree's other branches.

It is the totality of species—the vast range of life strategies and competencies—that constitutes the wonder and magnificence of life on Earth, not the particular attributes of any one species. In the end, all of us—fungi, fern, falcon, fish, and fox—are made of the same stuff—each species a unique unfolding of intelligences—all interacting and fitting together like pieces of a mysterious and dynamic tapestry.[29]

This profound revelation—that we are not *separate from*, but *participants in* life's web—coming from both science and the world's wisdom traditions, invites curiosity, calling us to see all of life's creatures as wondrous, each in their own way. I was reminded of this when, on a recent hike, I stopped by a stream to rest. As I sat with my feet dangling in the water, a tiny yellow spider dropped down in front of me on a silken thread. Delighted by her appearance, I enticed her onto my finger and looked at her with my hand lens. She was magnificent, delicate, almost translucent—with a faint red spot on her abdomen. As I watched, she played out a silk thread from her spinneret, dropped down a few inches, and then reeled herself back up, a lovely display of acrobatics. Looking at her again with my hand lens, I was able to see fine bristles on her legs and the complex articulation of her mandibles. Though her *umwelt*—her particular way of being in the world—was radically different from my own, she was remarkable and precious in her uniqueness.

After my spider friend left my finger and moved onto my jacket, I began to write about her in my journal, endeavoring to capture the joy and wonder of our time together. Later, as I was putting away my journal, I felt a tingling sensation behind my right ear and, instinctively, rubbed at it. The tingling sensation, it turned out, was caused by the yellow spider. It was too late; her remains were now smeared across my fingertips. Prior to this experience, killing a spider wouldn't have mattered so much to me, but this time, because I had experienced a measure of kinship with this spider while marveling at her beauty and grace, I felt remorse. I suppose that this is the difference between intellectually recognizing that *we are ALL part of the tapestry of life on Earth* and having the visceral experience of this truth.

Applications and Practices: Cultivating Curiosity and Connection

> When you really think about it, everything is connected—it's all connected.
>
> —Sally (Everglades woman)

Reading a book chapter like this one is not sufficient to create a genuine connection to nature, but words can serve as a springboard for relationship, provided we are willing to step a bit beyond our comfort zones.

Fifty Questions

One of my favorite practices for coming into relationship with nature is Fifty Questions. All that is required is a spot in a meadow or a forest or just a swath of lawn. Find your spot and sit quietly for a half hour, observing the life around you: the smells, the quality of the air, the sounds, the insects, the birds, the plants. As you attend to your surroundings, questions will inevitably arise: What bird is flitting about in that bush? Does she have a nest close by? What is that beetle carrying? Where is he going? Why is there moss on this rock but not on that rock over there?

Stay in the here and now, writing down your questions as they come to you. As you give yourself to this exercise, questions will surely arise—sometimes in torrents. You will also learn things about yourself, such as which of your senses—sight, sound, smell, touch, taste—tends to trigger most of your questions.

After a half hour, you easily will have accumulated fifty questions, perhaps even a hundred. Some of your questions will already have known answers, but it is likely that there will be some on your list that have never been asked by anybody, much less addressed.

Creating this list of questions is just a beginning. You don't have to stop there. You could become a citizen scientist by picking a question that is especially interesting to you and then combining your curiosity and common sense to seek an answer to your question.

Discovering Your Identity Through Clay

Who are you? Are you your accomplishments? Are you the bullet points on your résumé? Perhaps, yes, in part. But rather than relying on your powers of reason and discrimination (your so-called left-brain attributes) to answer this question, what if you relied on your visual, imaginative, sensual, and intuitive capacities (i.e., your right-brain attributes) as a gateway to self-discovery?

One way to stimulate your right brain is to direct the question, "Who am I?" to EARTH! This activity might sound far-fetched, but think about it: Each of us is constituted of Earth; we were each born out of her, and some day we will each return to her. So why not ask Earth?

To make this inquiry less abstract and more visceral, go out and get yourself a fistful of clay. You could purchase clay at a craft store, but in many parts of the United States all you need to do is simply dig down a foot or so into the ground and you will likely find a seam of rusty-red clay. Construction sites and creek banks are also good sources of clay.

Once you have some clay, take it into your hands, sensing its weight, texture, and smell. If necessary, add some water to make it more pliable. Then, close your eyes and begin to explore the clay with your moistened hands, noting the contours, depressions, and bumps, the wet and dry places, the warm and cool zones. With your eyes still closed, give yourself permission to play with the clay.

Leave your thinking behind, allowing your hands to take the lead, molding the clay in ways that simply feel good, right, enlivening. As you sink deeper into this creative process, say to yourself: "This clay is me and I am creating myself." There is no hurry. Allow your hands to decide when you are really done and only then open your eyes to behold your creation. Run your fingers and eyes over and around the clay form that you have created, both looking at it and touching it, attentive to any feelings that arise. If further sculpting is called for, have at it.

When you are finished, ask: What insights, longings, visions, and fears are embodied in my sculpture? What is this clay—shaped by me—telling me, revealing to me? As you reflect, take to heart these words from the famous psychologist Carl Jung: "Often the hands know how to solve a riddle with which the intellect has wrestled . . . in vain."[30]

Place your creation on a stand to dry. Look at it from time to time—for example, before you sleep and/or when you wake up. Trust that the body of Earth—your larger body—has things to teach you, a story to tell you, about who you really are.[31]

Questions for Reflection

- How would you describe your relationship with nature? How has this relationship changed over your lifetime?
- It has been said that we are all scientists—all curious, all creative, all problem solvers. How might this be true for you?
- How does the realization that your body is a swirling, pulsating ecosystem, providing habitat for millions of creatures, affect your identity?
- To what degree are you a *speciesist*—that is, someone who thinks of humans as superior to other species? What are the implications and consequences of this worldview?
- How might experiencing the *umwelt* of other organisms create greater intimacy—greater connection—between you and Earth's web of life?
- What happens for you when you allow yourself to consider that, contrary to appearances, there may be no such thing as a separate, autonomous self, insofar as the underlying quantum reality of existence suggests that everything is intimately entangled?

3

Intimacy

Belonging to Earth

> Tell me the story of the river and the valley and the streams and woodlands and wetlands, of shellfish and finfish. . . . A story of where we are and how we got here and the characters and roles that we play. Tell me a story . . . that will be my story as well as the story of everyone and everything about me . . . a story that brings us together under the arc of the great blue sky in the day and the starry heavens at night.
>
> —Thomas Berry[1]

What is Earth to you? A rock in space? A bundle of resources? A dumpsite for your waste? Many of us, living in the "modern world," see Earth in these utilitarian and soulless ways. But *what if* Earth isn't just a backdrop for our busy lives? Instead, what if we were to see Earth as our Mother, as our larger body, not metaphorically but literally? Think about it: Just as your liver cells and skin cells and muscle cells are part of your physical body, so, too, are you a part of Earth's body. Indeed, Earth is primary; she came first. Each of us is a derivative of Earth, made from and utterly dependent on her for our every breath.

How do you react to this? What would it be like for you to experience yourself as a part of Earth, as belonging to her? Is this just a farfetched idea for you, or do you receive it as a tantalizing possibility?

Foundation 3.1: We Belong to Earth

We didn't design our bodies, we inherited them. This inheritance is evident in our very hands. Pause for a moment to behold your own hands as if viewing a rare treasure. Inspect them, as an anthropologist might, noting their paw-like design. Explore your fingers as if seeing them for the first time, playfully moving them, delighting in their agility. Close your eyes and use one hand to investigate—through touch—the bone structure, joints, and musculature of your other hand.

Your hands are gifts from the universe. It is easy to take them for granted, but consider: Without your hands you would be compromised when it came to lifting, carrying, holding, grasping, squeezing, gesturing, reaching, touching, and nurturing.[2] It's taken tens of millions of years of conditions unique to Planet Earth to shape and hone the human hand into its present form. The same holds for every other feature of our bodies, from our feet and knees and hips to our hearts and eyes and ears. In each case, Earth's conditions and environments have shaped our human form. It's no different for Earth's other creatures, as writer David Abram observes:

> We humans are corporeally related, by direct and indirect webs of evolutionary affiliation, to every other organism that we encounter. . . . In a thoroughly palpable sense, we are born of this planet, our attentive bodies coevolved in rich and intimate rapport with the other bodily forms— animals, plants, mountains, rivers—that compose the shifting flesh of this breathing world.[3]

It is easy to take our bodies for granted, but in so doing, we deny ourselves opportunities to experience awe and wonder. You can create your own mind-blowing moments by slowing down to bring your attention to ordinary daily acts, like picking up an object with your hands or bringing water to your lips and swallowing. For example, I reacted with amazement when I learned that the seemingly ordinary act of *seeing* my friend's face requires tens of millions of neurons, involved in complicated feedback loops that harness distinct parts of my brain. Likewise, I was stunned when I learned that my liver is engaged in more than thirty thousand enzymatic reactions each second. In view of this, the esteemed scientist Lewis Thomas said that he would prefer to be given the controls of a 747 airplane, knowing nothing about how to fly, than to be put in charge of the functioning of his own liver.[4]

Awe Within an Atom

I am awestruck by how much of life is invisible to us. For example, as humans, we are unable to see the microscopic bacterial, fungal, and planktonic surfaces (see figure 3.1) where so many of life's exchanges take place. Meanwhile, the things that we do see are, in a sense, illusions. For example, when you look in the mirror, you see a reflection that is the result of light bouncing off the atoms that make up your face. Although you may look solid, you are mostly emptiness—visible emptiness! Cosmologist Brian Swimme explains it this way:

> You are more emptiness than you are created particles. We can see this by examining one of your atoms. If you take a single atom and make it as large as Yankee stadium, it would consist almost entirely of empty space. The

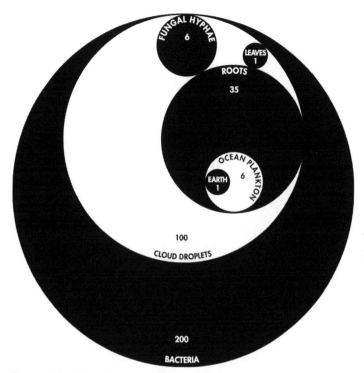

Figure 3.1 The surface area of Earth relative to other life surfaces.

center of the atom, the nucleus, would be smaller than a baseball sitting out in center field. The outer parts of the atom would be tiny gnats [electrons] buzzing about at an altitude higher than any pop fly Babe Ruth ever hit. And between the baseball and the gnats? Nothingness. All empty. You are more emptiness than anything else.[5]

The reason that the chair you sit in or the cup you drink from seems to be solid is that electromagnetic forces operate to hold the speck-like nuclei of atoms together. Upshot: Things are not as they seem! This realization can prompt humility and, with it, awe and wonder.

Life as Exchange

If I were asked to sum up *life's* operating systems using just one word, I would choose *exchange*. Whether human or frog, worm or bacterium, all of life's creatures are *designed* for the exchange of the minerals, gases, nutrients, and water that all of life requires. In the human case, the surfaces of our lungs receive Earth's air, and the surfaces of our gastrointestinal systems receive Earth's nutrients. Insofar as exchange surfaces are a basic design feature for all life, it's no wonder that all living things are mostly comprised of surfaces (see figure 3.1).[6]

If you are wondering where these surfaces are located, just step outside and take in a tree, noting its leaves. If all the leaves of Earth's land plants were stitched together, they would make a complete covering for the Earth; that is, the cumulative leaf surface area of all land plants is roughly the same as the surface area of Earth. This fails to account for all the microscopic plants (known as phytoplankton) that inhabit the Earth's oceans. Remarkably, the combined surface area of these microscopic plant forms is roughly six times greater than Earth's total surface area. This ratio may seem incorrect, especially because the total mass (think weight) of the phytoplankton in Earth's waters is small compared to the total mass of leaves on land, but appearances can be deceiving.

This paradox is resolved by the fact that—unlike land plants, with their large, flat multicellular leaves—ocean phytoplankton occur as separate cells, and this makes a difference when it comes to surface area. For example, imagine dispersing the hundreds of thousands of cells in a single tree leaf, one by one, into the ocean. This would increase the total surface area of that leaf dramatically, which is precisely the strategy of ocean phytoplankton—that is, to be small and numerous. If this is difficult to grasp, imagine that I gave you the task of measuring the surface area of a medium-sized log. Then, suppose that after you completed your task, I ground up that log, turning it into saw dust, and asked you to calculate the combined surface area of all the tiny particles of sawdust that previously had comprised the log. Before you even began, you would know that the cumulative surface area of all that sawdust would be many times greater than the external surface of the log that you measured initially.

THE PLANT-SUN PARTNERSHIP

You and I, along with all of Earth's other animals, are able to find food because Earth has two crucial components. Without these components we'd starve. If you are not sure what they are, look up and then look around. What are two things that you see that are essential for your survival? Answer: (1) look up and see the Sun and (2) look around and see plants everywhere.

Through the alchemy of photosynthesis, the plant world creates food using solar energy, and it is this food from the plant world that sustains (directly or indirectly) all creatures that walk, crawl, wriggle, run, fly, swim, hop, and burrow. Indeed, we are all alive and breathing thanks to the partnership forged between Earth plants and Earth's Sun![7]

And what about bacteria, the most abundant of all organisms on Earth? Their cumulative surface area exceeds that of Earth's by some two-hundred-fold (see figure 3.1). And let's not forget Earth's soils and water and air—the exchange surfaces that nurture life in all its forms. It turns out that the surface area of Earth's topsoil is more than ten thousand times the surface area of Earth herself; any way you cut it, water is all surface, and air is too.[8] Seeing the loci of life's processes as *surfaces* reminds us that the essence of life is exchange—giving and taking . . . relating . . . sharing.

Lifting Up and Flowing Down

Everything in the natural world can be thought of in terms of exchanges involving *lifting up* (growth) or *flowing down* (decay). The vigorously growing garden plants of summer are lifting up, while the leaves decaying on the ground in fall are flowing down. Likewise, the cow pie deposited in the pasture flows down, releasing nutrients to the soil that will eventually be lifted up in the form of new grass that will support cow growth.[9] Organisms classified as *decomposers* do the critical work of breaking down dead organic matter, ensuring that nitrogen and phosphorus and other vital nutrients flow to the soil, where they become available to be lifted up by plant roots.

In the early stages of decomposition, ants, worms, millipedes, beetles, and their counterparts play an important role by physically chewing apart the plant debris, greatly increasing the surface area available for later action by bacteria and fungi.[10] It is possible to witness, firsthand, the dynamics of all this recycling. You need only paw around a bit in the soil with a hand lens to gain a glimpse of these decomposers:

> These [creatures] are often in motion, hurrying along the vast expressways made by moles, the boulevards of earthworms, the alleys between particles of sand or clay. . . . Certain districts and certain intersections—mainly close to the roots of plants—get especially busy. The [creatures] move in the dark, sniffing out chemical trails. They are constantly doing business with one another. They traffic in molecules: minerals, organic compounds, packets of energy.[11]

Though it can be humbling to acknowledge, it is the little things—such as the bacteria that live in our guts and the tiny decomposers that populate Earth's soils—that literally *run the world*, overseeing the daily exchanges of nutrient capital necessary for planetary health. We've been conditioned to think that it's the government and the corporations, but from a biological perspective, it truly is the little things that run the world. They don't need us, but we sure as heck need them.

A WORLD WITHOUT DECOMPOSERS?

What do you suppose would happen if there were no such thing as decomposers? Think about it from the perspective of a forest. Without decomposers, all the leaves and branches and trees that die each year would simply accumulate on the forest floor; after a few decades, there would be so much dead debris on the ground that it would be difficult to even walk through the forest. Worse still, if there were no creatures around to set free the mineral elements locked in the dead leaves and downed trunks, the forest would eventually weaken and die for lack of nutrients. But of course this is not the way things work. Gazillions of decomposer organisms, dwelling on the forest floor, feed on the tree's dead organic remains, replenishing the soil with the essential nutrients that the forest needs to survive.

The Cycles of Life

Lying in bed one evening during a spring rainstorm, I experienced an abiding sense of peace as I listened to the rain settling onto the land. Why is it, I wondered, that I find it so comforting to be in the presence of rain? And then, suddenly, I knew: Rain—water—is what mostly makes up our bodies. What's more, water was the medium that brought us into existence. Indeed, in the most intimate of human acts, spermatozoa from our father were set free to swim up a canal in our mother toward an egg. Sperm joined with egg, and we began to grow, spending the first nine months of our lives in a salty sea of amniotic fluid (98 percent water).[12] Today, we are still mostly water. Each day we take in approximately two quarts of water, either through the liquids that we drink or as constituents in the foods that we consume. This water is essential to our survival, as it is only in the presence of water that the cells of our bodies are able to function.

Tracking an Atom

All of life's essential elements, including the hydrogen and oxygen in water, are constantly on the move, cycling up and down, in and out. One way to appreciate these cycles is to imagine what it would be like to track the meanderings of a single atom over time. You can do this by applying the *umwelt* concept introduced in chapter 2. Begin by imagining that you are a calcium atom encased in a dead log that has just been set free through the process of decomposition. Initially, you find yourself dissolved in a drop of water lodged between two grains of sand, but soon you are sucked up by a tree root and incorporated into a leaf. When winter sets in, the leaf falls to the ground. In the spring, microbes break the dead leaf down into bits of organic matter. Then an earthworm consumes that organic matter and after digestion, excretes you in the form of a worm casting. You are no sooner set free than you are lifted up, once again, by a tree root.

This cycle repeats many times—soil to tree and back to soil—until one day you find yourself encased in an acorn. On a frigid winter afternoon, a squirrel consumes the acorn and you become part of that squirrel's body. Several months later, the squirrel dies by the edge of a stream and is washed into the water during a spring flood. As you (a calcium atom) float down the stream, microbes set to work on the squirrel's carcass and once again you are set free—but not for long. An algal cell absorbs you into its tiny body, and this cell is consumed by an insect larva, which is then gobbled up by a trout. When the trout dies, you (calcium atom) are set free by decomposers as they dismantle the trout's body.

Decade after decade, you join one organism after another. Meanwhile, the inexorable force of gravity takes you further downstream until one day you reach the ocean. You continue to be shunted from organism to water to organism, until you eventually end up lodged in the bone of a sperm whale. When that whale dies, its bones settle to the ocean floor and, over millennia, become buried in sediment. It is possible that you could remain locked in ocean sediments for millions or even billions of years. However, a time may eventually come when the clashing of Earth's tectonic plates lifts the ocean sediments (of which *you* are a part) upward in a fit of mountain building, inviting you back into the cycles of exchange that are essential for the continuance of life.

A Nutrient-Cycling Case Study

Many scientists have dedicated their lives to understanding nutrient cycles insofar as these cycles are essential for a healthy planet. By way of example, some years back, two American ecologists, Herbert Bormann and Gene Likens at Yale University, initiated an ambitious nutrient-cycling study on a watershed in the mountains of New Hampshire. The stream running through this watershed—Hubbard Brook—collects all the water draining from the surrounding mountains.

Bormann and Likens conceptualized the watershed as a network of compartments or boxes, with trees in one compartment, soil in another, and so forth. The nutrients (e.g., potassium, nitrogen, phosphorus, calcium) in any given ecosystem box move among compartments, as shown by the arrows in figure 3.2.[13]

The starting point for their study was to determine the total amount of each nutrient in the various ecosystem compartments. Imagine the challenge: You walk out into the watershed to determine the amount of calcium in the forest-tree compartment of the ecosystem. How would you do it? The first challenge would be to figure out the total weight (mass) of the forest, and then you'd have to figure out the percentage of that mass that is com-

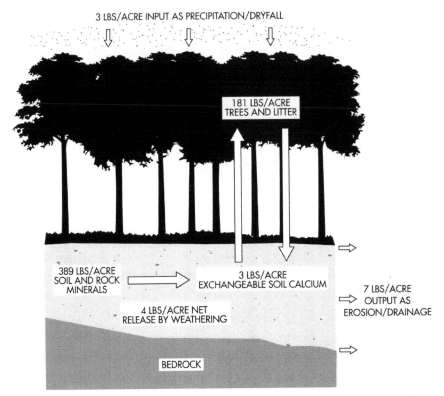

Figure 3.2 Schematic map of the Hubbard Brook ecosystem, showing the amounts of calcium in various forest compartments and the movements of calcium among compartments and into and out of the ecosystem.

prised of calcium. But weighing the whole forest, even if it could be done, would leave you with a huge pile of logs, branches, leaves, and roots; and, of course, you would end up destroying the very forest that you had hoped to study. The solution—rather than cutting down and weighing each tree in the forest—was to measure the diameter and height of all the trees and estimate their weight based on those measurements. This approach yielded a good estimate of the weight of the entire forest compartment. Next, the researchers took samples of all the tree-tissue types—leaves, bark, trunks, roots, and so on—and using prescribed chemical testing procedures determined the concentrations of calcium and other important nutrients in each tissue type. By scaling up, they were then able to estimate the total amount of each nutrient element in the aboveground forest vegetation compartment. Similar approaches were taken to estimate the nutrient stocks in other ecosystem compartments (e.g., soils, animals).

If you think this kind of work is tedious, I can affirm from my own experience that it is! And there is more. The researchers also had to determine the quantity of each nutrient entering the watershed (inputs) and leaving the watershed (outputs). Nutrients entered the watershed in rain and snow and as dust particulates (i.e., *dry fall*). Precipitation collectors were used to measure these atmospheric inputs, while a stream-monitoring station at the base of the watershed was used to monitor the quantity of water leaving the watershed. By conducting chemical analyses of both the precipitation coming into the forest and the stream water leaving the ecosystem, Bormann and Likens were able to calculate the quantities of nutrients entering and leaving the entire watershed.

So what's the value, the end result, of all this toil? Aside from revealing patterns of nutrient movements within forest ecosystems, the study showed that New England forests are very thrifty—that is, they don't waste much. For example, for each acre of forest in the Hubbard Brook watershed, only seven pounds of calcium were lost each year, whereas three pounds of calcium/acre/year entered the watershed via precipitation and dry fall. Hence, the net annual loss of calcium was only about four pounds/acre/year, a tiny amount compared to the almost six hundred pounds of calcium per acre in the soil/vegetation compartment (see figure 3.2). The researchers concluded that this forest, in its natural state, was extremely efficient, retaining 99 percent of its calcium from year to year.[14]

In forests everywhere, nutrients cycle efficiently as long as humans don't interfere with nutrient uptake (lifting up) and decomposition (flowing down) processes. However, when forests are significantly disturbed—for example, through careless logging practices—they begin to hemorrhage, leaking their nutrient stores, and when this loss occurs forest vigor is compromised.

Nutrient cycling isn't restricted to forests; it occurs everywhere, all the time. Every day each of our individual behaviors can enhance or disrupt Earth's nutrient cycles. For example, putting lawn leaves in plastic bags and sending them to a landfill (as many Americans do) isolates these valuable organic materials from natural nutrient recycling processes. This need not be. Another option is to keep household nutrients in circulation by composting leaves, grass clippings, and kitchen scraps and using the resultant compost to fertilize household gardens.

Foundation 3.2: Experiencing Ourselves in Life's Web

One of my most exciting and baffling experiences in the wild occurred while on a weeklong canoe trip on the Spanish River in Ontario with a lifelong friend. One afternoon, while having lunch by the river's edge, we noticed a large animal in the distance. At first, s/he looked like a horse, but we soon realized that it was a young cow moose that was trotting down the river toward us. We watched, in perfect stillness, as she approached: three hundred yards . . . two hundred yards. At a hundred yards, she disappeared into a meadow and then suddenly reappeared, bursting out of the undergrowth, just twenty-five yards upstream from where we sat, mesmerized. She was so close that we could hear her labored breathing and see welts from fly bites on her flanks. When she was just ten yards away, she crossed to the other side of the river on a gravel shoal and disappeared into an alder thicket.

As we sat wondering why this moose seemed so rattled, we sighted a lone white wolf trotting down the river toward us. Like the moose, the wolf also disappeared into the meadow. Anxious to follow her movements, we scampered up onto a ledge. In the very moment that we gained stable footing, that wolf appeared on that same ledge, just ten paces away. At the sight of us, s/he spun around and vanished into the underbrush. I don't know what happened after that; I do know that for a brief moment my friend and I were witness to a life-and-death drama.[15]

Stepping into Life's Web

It is not necessary to travel to a remote wilderness to have a direct encounter with the natural world. To see what I mean, I invite you take a walk with me. Begin by removing your shoes—those fabricated contrivances that separate your flesh from the *flesh* of Earth. Allow your extraordinary feet, those "down-turned hands at the end of your hind legs," to direct your movements.[16] Feel the coolness of Earth, the textures underfoot; hear the sounds, both soft and crackly, created with each footfall. Be alert to how this physical experience of the world underfoot awakens you to the world around you.

There is more going on here than meets your eyes. For example, as your feet touch and sense Earth's *skin*, there are millions of organisms within the soil that are sensing your presence. Yes, wherever we walk, the microscopic creatures underfoot *sense*, in their own ways, your presence.

Remind yourself that in this moment you are reclaiming part of your birthright. The word *human*, after all, has its origins in *humus* or soil; just as the word *matter* has its roots in *mater*, or mother. So it is that, with your bare feet, you are in direct contact with the "mater" (mother) that nurtures and sustains you. Notice, too, how with each step, Earth is there to receive you.

When the time feels right, find a seat on the ground—that is, on the dirt that is Earth's skin. In so doing, you will be reenacting what your human ancestors did for thousands of generations. After all, sitting in chairs and sleeping in beds is a very recent practice for humans—one that, while offering certain comforts, separates us from direct contact with Earth. So, with the innocence and inquisitiveness of a child, explore what it is like to have your body in direct contact with the body of Earth.[17]

As you surrender to the ground, notice, too, that you are not so much breathing as *being breathed*. Breathing happens involuntarily, about sixteen times each minute. With both humility and gratitude, acknowledge that it is not your air but Earth's air that you are receiving.

Imagine, now, that your attention is captured by a mosquito bobbing up and down in front of you. She has located you by following the plume of carbon dioxide released in your exhalations. When she alights on the back of your left hand, your first impulse is to squash her, but your curiosity overrides your annoyance and you watch, with macabre fascination, as this mosquito's abdomen fills with blood—your blood! You even manage to smile a bit, acknowledging that by giving her some of your blood, you are participating in nature's *gift economy*.[18] In this vein, David Abram reminds us, "No matter how earnestly we humans strive to exempt ourselves . . . we cannot escape our participation in [nature's] cycles of exchange. If we ingest the land's nourishment . . . incorporating the world's flesh into our own—it can only be so because we, too, are edible. Because we, too, are food."[19]

Trees Versus Us

As humans, it is pretty easy for us to specify all the ways that we are different from insects and birds and worms and fish and crabs and plants, but we struggle when it comes to articulating the ways that we are similar to these other life forms. For example, right now, can you come up with five things that you share in common with trees? Is this easy for you to do? Or is your mind more fixated on all the ways that you are different from trees—for example, you can walk, trees can't; you have a brain, trees don't; you can talk, trees can't . . . and so forth.

But what if we actually have a lot more in common with trees than we realize? Before dismissing this as nonsense, consider the following eight similarities between trees and you:

1. *Elemental composition*. No part of our human anatomy looks anything like a plant leaf, yet when we compare the ratios of nitrogen, carbon, phosphorus, and other vital elements in plant leaves to these same elements in our bodies, the ratios are remarkably similar. Said differently, the elemental composition of the greens on your plate matches up with the elemental composition of the tissues of your body.[20]
2. *Biochemistry/genetics*. Tree tissues, like human tissues, are composed of carbohydrates, fats, and proteins. Plants use the same amino acid building blocks to construct the same proteins that we construct. They also employ very similar, sometimes identical, biochemical pathways for the synthesis and breakdown of compounds in their cells. Like humans, plant cells contain a nucleus with DNA packaged in chromosomes. In addition,

plants have an array of hormones that act as messengers, turning cellular processes on and off, just like us.[21]

3. *Sex.* It's not just humans and other animals that engage in sex; plants also have a sex life. Flowers are the genitalia of trees and all other flowering plants. The minuscule pollen grain is analogous to the male genitalia. When a pollen grain arrives at a flower (transported on the wind or carried by an insect), it germinates and grows a penis-like tube down into the flower's ovary, where sperm is released to fertilize an egg.

4. *Population expansion.* Successful mating in humans and trees leads to the production of offspring (seeds/seedlings, in the case of trees). Although trees are stationary during their adult lives, their seeds can be carried considerable distances by wind, water, animals, and other means. Because of this, trees, like humans, are able to colonize new territories.

5. *Circulatory system.* Just as humans have a circulatory system, composed of arteries and veins, to transport nutrients from place to place throughout our bodies, the same is true for trees. In their case, one set of vessels carries water and nutrients up from the roots to the tree's canopy and another set takes sugars, created in leaves, down to the roots for nourishment.

6. *Defenses.* Without some sort of defense system, many trees would be attacked by insects. Plants avoid this fate by creating chemicals that are repulsive to insects. Many of these chemicals simply taste bad; others interfere with digestion. All send the message: "Stay away!"

7. *Communication.* Just as humans send out health warnings to each other and take precautionary measures (e.g., getting a flu shot at the beginning of flu season), many plant species also send warnings to each other when there is danger afoot. For example, if the leaves of an individual maple tree are being attacked by caterpillars, that tree can respond by releasing an airborne chemical message to other maples nearby, saying, in effect: "Hey neighbor, I am getting eaten. If you want to stay healthy, you better create some bug-repellent chemicals right now."

8. *Partnerships with microbes.* Humans are not the only ones who depend on a multitude of microbial partners (see chapter 2); the same is true for trees. For example, oaks, pines, maples, and hickories, among many tree groupings, have specific species of fungi that associate with their roots. The tree provides food to these fungi, and the fungi reciprocate by sending out a network of fungal threads from the tree's roots, thereby dramatically improving the tree's ability to absorb essential nutrients from the soil.

Interlude: Intimacy with Paper

Take a break from reading now and simply examine the paper that this book's words are printed on. What do you see? What is *paper*? If you examine this page with a magnifying glass, you will see that it is composed of fibers (think vessels) that were once part of a tree's circulatory system. Yes, in a literal sense, paper is a tree's circulatory system rendered flat.

There are other, more metaphysical ways to capture the essence of paper. For example, Buddhist monk Thich Nhat Hanh captures the essence of paper this way:

> If you are a poet, you will see clearly that there is a cloud floating in this sheet of paper. Without a cloud, there will be no rain; without rain, the trees cannot grow; and without trees, we cannot make paper. . . . If we look into this sheet of paper even more deeply, we can see the sunshine in it. Without sunshine, the forest cannot grow. . . . And if we continue to look, we can see the logger who cut the tree and brought it to the mill to be transformed into paper. And we see wheat. We know that the logger cannot exist without his daily bread, and, therefore, the wheat that became his bread is also in this sheet of paper. The logger's father and mother are in it too. When we look in this way, we see that without all of these things, this sheet of paper cannot exist.[22]

How we perceive the world and our place in it determines, to a significant degree, our actions. If a tree is simply a resource—a commodity to be harvested—then we will objectify *it*. If, on the other hand, we perceive trees as beings with their own unique intelligences and integrity, we will naturally be inclined to treat them with respect and perhaps even affection. In this, I take inspiration from English professor Scott Russell Sanders, who wrote: "I confess that I do hug trees in my backyard and any place else where I happen to meet impressive ones. I hum beside creeks, hoot back at owls, lick rocks, smell flowers, rub my hands over the grain in wood. I'm well aware that such behavior makes me seem weird in the eyes of people who've become disconnected from the earth. But in the long evolutionary perspective, they're the anomaly. Our bodies were made for this glorious planet, tuned to its every sound and shape."[23]

Terraforming: A Thought Experiment

As conditions on Planet Earth continue to deteriorate, some scientists believe that the time has come to develop strategies for colonizing other planets, thereby ensuring humanity's long-term survival. Noting this trend, Stanford biologist Gretchen Daily developed a challenging thought experiment.[24]

It begins by imagining that you have been put in charge of an expedition that will create an independent, self-sustaining civilization on the Moon. To make the challenge a bit less daunting, assume, first, that the Moon has a friendly atmosphere, just like our own, with lots of oxygen; second, that the Moon's surface is composed of pulverized rock (the basis for soil); and third, that the Moon has lots of water. Your job is to decide which organisms you will need to bring from Earth to ensure the successful establishment of a permanent human colony on the Moon. The expedition commander assures you that your spaceship is massive in size so that you will have ample room to accommodate whatever you need to carry with you.

How would you proceed? Which of Earth's organisms would you take with you?

Presumably, food is the first thing that comes to mind. Though most of us associate food with packaged items that we purchase in grocery stores, food comes from Earth's plants and animals. Recognizing this basic truth, the first thing on your list is seeds of grains (corn, wheat, oats, rice, barley, etc.), vegetables (carrots, onions, kale, potatoes, squash), fruits (bananas, oranges, apples, grapes, kiwis, mangoes), and herbs or spices (thyme, rosemary, black pepper, ginger, garlic). Then, of course, you will need seeds to grow coffee and tea, as well as hops (for beer). And oh yes, you will also want cane plants for sugar and cacao plants for chocolate. And, finally, you mustn't forget to bring along spores from an array of edible mushroom species.

In the process of making your food list and gathering the necessary seeds, you realize, perhaps as never before, that humans rely on hundreds of different species of plants. And not just plants! If your mission is to establish a sustainable civilization on the Moon, it would be prudent to also take animals with you. So, to your list you add chickens, goats, cattle, sheep, and turkeys, as well as water tanks filled with trout, salmon, shrimp, flounder, clams, oysters, and so on.

Just when you think that your job is done, you remember that you will need to make clothes, so you add cotton and hemp plants to your list and perhaps silk-producing caterpillars along with plant species that will provide dyes. You will also need to build things from wood and to create paper, so onto your list go dozens of different tree species.

"There, that should do it," you say to yourself, but then you remember that you will have to feed all your goats, chickens, cattle, and sheep. So you add seeds of alfalfa, forage grasses, nutritious herbs, and hardy edible shrubs to your list.

At this point, you are pretty sure that you have everything covered, but to be certain, you picture yourself actually living on the Moon. There you are with all your seeds and animals; but as you go to plant them, you realize that the soil is sterile—barren of the bacteria and fungi that are essential for plant nutrition and nutrient cycling. So you collect thousands of tons of living soil to bring along on your expedition, in hopes that the organisms living in the soil will reproduce and spread.

As you imagine your crops and fruit trees growing, you realize, with a jolt, that you forgot the insects, birds, and bats that will be necessary to pollinate all the food plants, so you set about compiling a list of the pollinator species that you will need. In so doing, you begin to wonder what your various pollinators will survive on during those times when their preferred plants are not supplying pollen and nectar. This concern sends you in search of wild (noncrop) plant species that will supply the various pollinators with food throughout the entire lunar year.

Then one night, just a month before your scheduled departure, you awaken from a nightmare. In your dream, you are newly settled on the Moon and insect pests are destroying your crop plants. This reminds you that the domestic animals and plants that you will be taking to the Moon will undoubtedly harbor microbial and insect pests. Since synthetic pesticides are not effective in the long run, you will need to take along the natural enemies of all your crop and livestock pests.

Finally, as you really think hard for the long term, you realize that, given enough time, environmental conditions (e.g., climate) will change on the Moon, just as is the case on Earth. Hence, it won't work to take just a single variety of each species of plant and animal. Instead, you will need to have as many strains as possible to ensure that you have the raw material (genetic variation) for evolution. Upshot: This thought experiment highlights that, as humans, we are deeply dependent on Earth's biodiversity in *all* its forms.

Lessons from Biosphere 2

It is sobering to acknowledge that, in spite of all our sophisticated knowledge, we lack the scientific know-how to create self-sustaining ecosystems here on Earth (much less on the Moon). Our shortcomings in this regard were illustrated by the failure of the Biosphere 2 experiment.[25]

Biosphere 2 was designed to function as a completely self-contained, artificial ecosystem. It was constructed under a transparent, airtight bubble in the Arizona desert. Three acres in size, this $200 million plus unit featured zones of agricultural land, rain forest, desert, and savanna, along with wetlands and a tiny ocean with a coral reef. Biologists were hired as consultants to design the various subsystems and to determine the array of plants, animals, and microbes necessary to sustain life in this micro-biosphere.

After everything was set, eight humans were sealed into the unit. The team intended to stay for two years, which they did . . . barely. The experiment yielded many interesting results, including the fact that the air quality in the unit deteriorated. Specifically, oxygen concentrations dropped from a normal 21 percent to 14 percent, an oxygen level typical of elevations of thirteen thousand feet, and nitrous oxide concentrations rose to levels that can impair brain function. At the same time, nineteen of twenty-five vertebrate species in the unit went extinct along with all the pollinators, thereby dooming most of the plant species to eventual extinction. In addition, algal mats polluted the water, while ants and cockroaches experienced population explosions.

The eight Biospherians and all of the scientists behind the scenes learned a basic lesson the hard way: We, humans, are deeply insinuated into the living body of Earth, and our knowledge and sophistication pale in comparison to the complexity and innate sophistication of nature.

Stepping Back to See the Big Picture

Rather than responding to the defilement of Earth by scheming and dreaming about escaping to the Moon (or some planet in another solar system thousands of light years away), what if we were to commit to cultivating a deep and respectful relationship with our home places right here on Earth?

Maybe you think you already know your home place. For example, when someone asks you where you're from, you respond with the name of your state and town and a few other identifiers, but you may not have much to say when it comes to describing the physical and ecological characteristics of your home place—for example, the lay of the land, the geological features, the common trees and birds and insects, the soil types, the weather patterns, the smells and seasonal sounds that are unique to your home place. Brian Swimme has a simple test for assessing our degree of intimacy with our home places:

> It's easy to do. You simply invite someone to visit you who lives at least twenty miles away and who has never visited you before. You can give verbal instructions on how to get to your abode over the telephone, but the one rule is this: In your directions, you may refer to *anything but* human artifice [i.e., it is not fair to refer to man-made things such as roads, signs, houses, businesses]. Your job, then, is to guide your friend to your house by referring only to elements in the natural world. So it is that you may refer to hills, oak trees, the constellations of the night sky, lakes or ocean shores or caves, the positions of the planets or any ponds, trails, or prairies, the Sun and Moon, cliffs, plateaus, waterfalls, hillocks, estuaries, bluffs, woodlands, inlets, forests, creeks, swamps, bayous, groves and so on. Whenever your friend gets stuck, she is free to phone you for more directions, but the rule for her is that she must describe her location, again, without referring to any human artifice.[26]

Very few people alive in the United States today could pass this test, though I suspect that it would have been effortless for the Native peoples living here prior to European settlement.

It seems that our separation from nature grows more pronounced with each generation, in part because of fear. For example, since the 1970s the radius around the home where children are permitted to roam has shrunk by more than 80 percent. Rather than frolicking outside in nature, children today often remain inside, glued to their screens; for example, kids between the ages six and fifteen now spend an average of six and one-half hours a day in front of screens.[27]

Upshot: Our children are being raised in a social environment unlike any other in human history. Eco-psychologist Chellis Glendinning posits that modern humans are undergoing, unbeknown to themselves, a profound form of traumatization. The *Diagnostic and Statistical Manual of Mental Disorders* defines *traumatization* as "an event outside the range of human experience that would be markedly distressing to almost anyone." In Glendinning's view, "The trauma endured by technological people, like ourselves, is the systemic and systematic removal of our lives from the natural world: from the tendrils and earthly textures, from the rhythms of sun and moon, from the spirits of the bears and trees, from the life force itself."[28]

A growing body of research now links the emotional and physical well-being of both children and adults to regular contact with nature. For example, hospital studies reveal that patients in rooms with a view of the natural world tend to do better than those without access to such a view. Similarly, researchers at Cornell University found that the more contact children had with nature—including such simple things as indoor plants and window views of nature—the better their ability to cope with stress.[29] These and other studies make the case that direct contact with the natural world reduces stress and improves well-being. Meanwhile, separation from the natural world—passing our time indoors—appears to have the opposite effect.

The possible calming effects of nature are of particular interest to researchers studying ADD (attention deficit disorder) among children and adults. A significant factor in the uptick of ADD in recent decades is likely the growing separation between children and nature, compounded by excessive electronic stimulation. To the extent that this is true, author Richard Louv suggests that *attention deficit disorder* might aptly be renamed *nature deficit disorder*—a condition resulting from the increasing alienation of humans from nature.[30]

It's not too late for all of us—adolescent kids, college kids, adult *kids*—to cultivate intimacy with the wild Earth that has birthed us into being. Relationship begins as we summon the courage to leave our indoor lives behind (along with our electronic gadgets) and step with curiosity and humility into the juicy shapes, sounds, smells, colors, textures, and movements of nature.

Wrap-Up: Expanding Ecological Consciousness

To awaken we must acknowledge the insidious effects of the objectification of Earth. Indeed, most of us have been conditioned to perceive Earth as an object that exists for our use. I recently observed this objectification of nature in the form of a welcome sign at the entrance to a national forest. The sign read, "Welcome: Land of Many Uses." You've probably seen versions of this sign. On the surface, the sentiment seems friendly, but the phrase—*Land of Many Uses*—is a declaration that the natural world is here for our use, that we are *top dog*, that it's all about us. But what if, upon seeing this sign, one of us had the audacity to cross out the word "Uses" and replace it with the word "Beings." A sign reading, "Welcome: Land of Many Beings" would be an invitation to experience both the forest and Earth, as a whole, as precious and wondrous and worthy of our deep respect and care.

Applications and Practices: Intimacy with Earth

> The proper goal of education is understanding what it means to be a human in a living world.
>
> —Stan Rowe[31]

Earth breathes, and we breathe with her. Earth has a metabolism, and we are part of her metabolism. Earth's elements cycle, and we are part of these cycles.

Intimacy with Earth's Breath

What could the phrase "joining to Earth's breath" possibly mean? If you'd like to find out, here's a simple practice. It's called *breathing with a tree*. Begin by respectfully approaching a tree, beholding her with curiosity, observing her trunk, her bark, her branches, and finally, her leaves. Then, focus on a single leafy branch, and as you do so, bring your attention to your breath, inhaling and exhaling slowly as you tune into the tree's gentle presence.

As you breathe out, remind yourself that carbon dioxide is leaving your body and passing into the tree's leaves. This means that some of the carbon that was within you, just seconds ago, is now being forged into sugars within the tree's leaves. The tree needs your carbon; it is a gift that you are offering her. The tree reciprocates by offering you her oxygen, a by-product of photosynthesis. As you breathe in, do so knowing that the tree's oxygen—her gift to you—is entering your lungs and oxygenating the cells of your body. By engaging in this practice, you give yourself the opportunity to participate, with full consciousness, in Earth's breathing.[32]

Intimacy with Earth's Body

Each time we eat, we have an opportunity to cultivate intimacy with Earth's body. One way to do this is to bring awareness to the act of eating. Physician Jon Kabat-Zinn suggests the following exercise to help his patients (people who have been living under chronic stress) eat mindfully:

> We give everybody three raisins and we eat them one at a time, paying attention to what we are actually doing and experiencing from moment to moment. . . . First, we bring our attention to seeing a raisin, observing it carefully as if we have never seen one before. We feel its texture between

our fingers and notice its colors and surfaces. . . . We note any thoughts and feelings of liking or disliking raisins if they come up while we are looking at [the raisin]. We then smell it for a while and finally, with awareness, we bring the raisin to our lips, being aware of our arm moving our hand to position the raisin correctly and of salivating as our mind and body antici- pate eating. The process continues as we take the raisin into our mouth and chew it slowly, experiencing the actual taste of one raisin. And when we feel ready to swallow, we watch the impulse to swallow as it comes up, so that even the swallow is experienced consciously. We even imagine, or sense, that our bodies are now one raisin heavier.[33]

No matter what we eat—whether it's a raisin, a carrot, a shrimp, a peanut, or a pretzel—it comes from Earth. This means that each time we eat we have an opportunity to cultivate intimacy with Earth, provided we remember that we are, literally, taking Earth into our bodies.

Giving Our Body to Earth's Body

Have you ever thought about what will happen to your body when you die? If you choose to follow the conventional American burial script, you will be embalmed with a mixture of chemicals (including formaldehyde and ethanol) and then placed in a metal casket that will be set in a concrete vault in a cem- etery. This combination of practices will effectively separate your biological remains from Earth's cycles. But it is possible, in most states, to be buried in a simple pine box, creating the conditions whereby your body's elements (nutrients) can be lifted up and recycled to join the great chain of life.

Steve VanMatre tells a story about recycling that involved a man, buried in a Massachusetts town, who had to be exhumed some years after his burial. Remarkably, when the man's remains were exposed, it was revealed that the roots of an apple tree had taken on the form of his body from head to toe.[34] Here was a man who, in a sense, had been reincarnated as an apple tree.

Cremation is another option that can reunite us with Earth's cycles. When my father died, my family took the elements from his cremation to the edge of a small pond that was teeming with life. We lifted his ashes up to the sky and then let them rain down upon the water. I remember being sur- prised that they sparkled in the sunlight. Before leaving, noting a blueberry bush by the edge of the pond, we spread the remainder of his ashes (nutri- ents) around its base. Then, we each ate a blueberry, as a reminder that each of us participates in the cycles of life and death each moment of our lives.

Questions for Reflection

- This chapter invites you to see Earth as your larger body. What happens when you do this?

- It's been said that "trees are our elders." What do you make of this?
- In what ways might you have lost your essential wildness and become domesticated? What do you see as the possible consequence(s) of this loss?
- Consider that, in biophysical terms, you *are* mostly empty space: How does this way of perceiving yourself affect your identity?
- What do you do in your daily life that disrupts or interferes with Earth's cycles, and what might you do to avoid such actions?
- If you engaged in the "breathing with a tree" practice, how was that experience for you? And if you were reluctant to try this practice, what do you make of this?

PART II
Assessing the Health of Earth

Unless we change the direction in which we are headed, we might
wind up where we are going.

—Chinese proverb

If I were to tell you that someone you don't know and will never meet was dying, chances are you wouldn't care very much. And that's probably a good thing. After all, one person dies every twelve seconds in the United States. That's five people every minute.[1] If we became upset every time one of these strangers died, we wouldn't be able to function.

But what if you had a personal relationship with one of these people? What if you depended on her for your sustenance and well-being? What then? This isn't a hypothetical scenario but instead the reality in which each of us now lives.

Who is this ailing being on whom we each depend? For most of us, she is so seamlessly integrated into our daily lives that we are unaware of how she cares for us—sustaining us with air and water and food—day by day. She *is*, of course, our Mother, Earth.

The thesis of part II of this book is that our Mother—this marvelous planet that has birthed us, nurtures us, shelters us, breathes us—is terribly sick and has been ailing for more than a century.

The deteriorating health of Earth was called to international attention in a 1992 statement issued by 102 Nobel laureates in science, along with sixteen hundred other distinguished scientists from seventy countries. That statement concluded with these words:

> We the undersigned, senior members of the world's scientific community, hereby warn all humanity of what lies ahead. . . . A great change in the stewardship of Earth and the life on it is required, if vast human misery is to be avoided and our global home on this planet is not to be irretrievably mutilated.[2]

Things have only gotten worse since 1992. Atmospheric chemists continue to report steady rises in greenhouse gases; agronomists remind us that soils are eroding more rapidly than they are forming; human physiologists report increasing concentrations of foreign, disease-inducing chemicals in our bodies; ecologists are registering the greatest spike in species extinction since the age of the dinosaurs; sociologists lament the breakdown of families and the dissolution of communities; while philosophers and religious leaders continue to call attention to the erosion of moral principles and the increasing alienation among humans.

Planetary breakdown is the unifying concept knitting together the three chapters of part II. In each of these chapters we examine Earth as a medical doctor might: first pinpointing signs of ill health and then looking below the symptoms to the underlying causes of breakdown.

4

Listening

Gauging the Health of Earth

> When you talk, you are only repeating what you already know.
> But if you listen, you may learn something new.
>
> —Dalai Lama[1]

I live in Central Pennsylvania, close to Spring Creek, a cold-water trout stream. One afternoon shortly after taking a faculty position at Penn State, I took a hike from my office to Spring Creek in hopes of visiting two of the freshwater springs that feed this creek. To reach the first one, Thompson Spring, I had to cross a busy highway, scale a chain-link fence, and then circle around a stinky wastewater treatment plant. After clawing my way through a tangle of briars, I encountered Thompson Spring, a lovely crystalline pool. Though the pure water bubbling from this spring was headed for Spring Creek, it got mixed with the dirty effluent from the wastewater treatment plant before reaching the creek.

Continuing my walk, I eventually reached Thorton Spring, emerging from the base of a forested hillside. This would have been a fine place for a picnic if it wasn't for the acrid smell. Thorton Spring, it turned out, was located just a short distance below a chemical plant that for many years stored its wastes in underground storage drums. Over time those drums developed leaks, and this resulted in the contamination of Thorton Spring (and by extension Spring Creek) with mirex and kepone, two nasty chemicals that eventually made their way into the bodies of the trout inhabiting the creek, as well as the bodies of the humans eating those trout.[2]

This all-too-familiar story underscores the fundamental ecological truth that *everything is connected*. Translation: If we make a mess in one place, the consequences will show up in other places. Over the past two centuries, the repercussions of our messes have been showing up everywhere—from the poles to the tropics—in the form of toxin-laden waters, polluted skies, contaminated soils, decimated species, climate chaos . . . the list goes on.

But maybe all this fretting over the state of the environment is exaggerated? Sure, there are some problems, but isn't this just the price we pay for progress? How do you see it? Are we on a slippery slope headed toward environmental collapse, or is everything pretty much OK when it comes to the health of the environment? This question has had environmental scientists' attention since the 1950s. Like medical doctors with a stethoscope to a patient's chest, they have had their ears tuned to planet Earth, listening attentively for signs of well-being, as well as danger.

Foundation 4.1: Listening to Earth's Sky and Land Creatures

Suppose that when you were a child there was a striking bird species—say Baltimore orioles—that nested in your neighborhood. But now, as an adult, when you return home, you no longer hear or see Baltimore orioles, and you wonder what has happened to them. You hypothesize that the environment of your neighborhood must have changed in some way that now makes it difficult for Baltimore orioles to survive. The idea that birds can act as indicators of environmental health is not new. U.S. coal miners used to take canaries with them into the mines, knowing that if the canary stopped singing (or worse, died), they must evacuate immediately because the concentrations of carbon monoxide had reached the danger zone.

The Situation Today

Since the middle of the last century, U.S. scientists in conjunction with everyday citizens have been using birds to gauge the overall health of Earth. Migratory bird species are ideal for this monitoring because they dwell in a wide range of environments, spending spring through early fall in North America (where they breed) and winter in warmer regions like Central and South America.[3] Examples of migratory bird species in the Americas are red-eyed vireos, wood thrushes, scarlet tanagers, red-winged blackbirds, ovenbirds, and Baltimore orioles, along with many species of warblers, hawks, eagles, and ocean shorebirds.

Collectively, migratory bird species dwell in a broad spectrum of environments. For example, some species prefer forests, while others favor grasslands, marshes, shorelines, and so forth. Scientists reason that if bird species, in general, are thriving, then it is likely that the habitats and environments that they depend on are also healthy.

Researchers employ several tactics to assess the health status of migrant bird populations. One straightforward approach is to census birds in the same parcel of land year after year. For example, the migratory birds breeding in Rock Creek Park in Washington, DC, have been surveyed each spring for more than a half century. The census is conducted by people who walk

WHY MIGRATE?

From a human perspective, it's logical to conclude that many species of North American birds migrate south each winter to escape the cold, but there's more to it than that. After all, if these birds go south for warmth, why do they bother to come back north in the spring? Why not just stay where it's warm rather than exert energy flying back and forth? The answer is food. By returning north in the spring, migratory birds gain access to the abundance of high-protein food that they need for the rearing of their young. Indeed, with the onset of spring, insects hatch out by the gazillions in the north. These insect hatches are timed to correspond with the unfurling of new tree leaves—an important food source for insect larvae.

Imagine for a moment (in the spirit of *umwelt*) that you are a bird arriving to a patch of forest with thousands of insect larvae feeding on every tree! What bliss! Now, imagine that you are one of those trees, with freshly hatched caterpillars beginning to munch on your unfurling leaves. What terror! How grateful you would feel for those migrant birds, plucking off the insect larvae that are eating your "body." Of course, trees don't experience "terror" or "gratitude," nor do birds (as far as we know) experience "bliss," but certainly reciprocity is a fundamental necessity for species coexistence within ecosystems.

through this large woodland park, day after day, identifying the birds that are present. Although this may sound tedious, it is exciting to be out early on a spring morning, listening to birds sing and observing their courtship and nesting behaviors. However, in the case of the bird count at Rock Creek Park, the enjoyment has been diminished because the census takers have discovered that the number of migratory bird species breeding in the park has dropped by one-third in recent decades.[4]

Of course, it could be that those migratory species that have disappeared from this park are still thriving in other areas of the United States. It would be necessary to have lots of Rock Creek–type studies, spread across the United States, before firm conclusions could be drawn. Fortunately, owing to an ambitious project initiated by the U.S. Fish and Wildlife Service in the late 1960s, just such a broad-scale survey exists.[5]

Each spring professional biologists along with competent citizen scientists conduct bird counts along more than four thousand prescribed transects (each one several dozen miles long) distributed throughout the United States and Canada. The results from this enormous undertaking, in tandem with research conducted by Cornell University, reinforce the Rock Creek findings, revealing that close to half of North America's forest-dwelling migratory bird species have experienced significant population declines in recent decades.[6]

Declines: A Closer Look

Documenting declines in the populations of migratory bird species is only one part of the research challenge; the other is explaining the causes of these declines. One obvious place to look for causes has been the tropics—the wintering ground for many of North America's migratory birds. We have all heard about tropical deforestation, and research has shown that a connection exists between the depletion of migratory bird populations and the depletion of their forest habitats in Central and South America, but tropical deforestation is only part of the story. Birds also face threats in the United States. For example, North American field biologists report that many migrant species are now having a difficult time raising their young. The birds return to the north in the spring, find mates, and lay eggs, but the breeding pairs often fail to fledge offspring.

Some of the factors at play may be evident in your own backyard. For example, have you ever seen a cat stealthily stalking a bird? I have on several occasions, but I never thought much about it until I learned that there are more than a hundred million *outdoor* cats (both domestic and feral) in the United States. Collectively, these cats kill between 1.4 to 3.7 billion birds each year—many of them migratory species.[7] Feral cats have proliferated, especially in the open, suburban-type environments that have been rapidly spreading throughout North America since the 1950s. Raccoons and opossums are also known to prey on bird eggs and, like outdoor cats, their populations have increased with the spread of suburbia.

Ecologist Dave Wilcove—noting that the broad expanses of forest that once covered many regions of North America have been sliced and diced into a mosaic of forest fragments—wondered if forest fragmentation could be a factor in the decline of migratory birds. Specifically, Wilcove hypothesized that migratory birds, nesting in small patches of forest surrounded by open land (e.g., suburbs), would be more likely to suffer nest egg predation by animals such as raccoons, opossums, and feral cats than migrants nesting in large, uninterrupted expanses of forest where feral cats, opossums, and raccoons are less common.

Wilcove tested his hypothesis by placing wild bird eggs in artificial nests and scattering these nests in fragments of forest ranging in size from a few acres to several thousand acres. He used the Great Smoky Mountain National Park, with five hundred thousand acres of contiguous forest, to represent the natural condition before the onset of forest fragmentation in North America. When Wilcove went back to check all the nests, he found that only one nest in fifty had its eggs eaten by nest predators in the Smokey Mountain site, whereas in the smaller forest patches, located in rural and suburban environments, egg predation was much higher, rising to 100 percent in some of the smallest forest fragments. This study offers strong evidence for a causal link between forest fragmentation and reproductive failure in forest-nesting migratory birds.[8]

This saga of migratory bird decline is also playing out in Europe and Africa. More than half of the hundred-plus migratory bird species that migrate between Europe and Africa each year are now experiencing population declines.

As if things aren't tough enough, there is now clear evidence that climate warming can be a further impediment to successful breeding for migratory birds. For example, spring is now arriving several weeks earlier in some areas of the United States and Europe, and this seasonal shift is causing insects to emerge ahead of the spring return of migratory birds, meaning that the birds miss out on a food resource important for rearing their young.[9]

In sum, populations of many migratory bird species, worldwide, are in decline. Are cats to blame? How about raccoons? What about tropical deforestation? Or should we point the finger at the land speculators and developers who are slicing apart forests with roads, malls, and sprawling housing developments? And don't forget the burning of fossil fuels that contributes to global warming and corresponding seasonal shifts in insect hatches. In fact, these are *all* factors that are contributing, in some measure, to declines in migratory bird populations. There is no single *smoking gun*, but it bears noting that the common denominator behind all these affronts to the welfare of birds is US! Yes, human activities on a multitude of fronts are jeopardizing the well-being of Earth's migratory birds.

It's Not Just Birds That Are in Trouble

If you had walked into an elementary school classroom in the 1950s and called out the word *extinction* and then asked students for the first thing that came to mind, chances are you would have heard just one word—dinosaurs! At that time, extinction was mostly understood as something that had happened far back in time. But call out *extinction* in a classroom today and students—in addition to calling out dinosaurs—are likely to volunteer the names of a raft of extinct, rare, and endangered species: passenger pigeon, California condor, blue whale, snow leopard, Florida panther, monk seal, mountain gorilla, and so on.

Biologists concur that habitat destruction is having severe impacts on the well-being of wild plant and animal species. Whether it is by replacing huge tracts of virgin rain forest in the Amazon basin with cattle pastures or building sprawling subdivisions in New Jersey where forests once stood or damming rivers in the Pacific Northwest, the outcome is the same: habitats—that is, the homes for millions of plant and animal species—are being destroyed at an unprecedented rate.

When scientists talk about the magnitude of the current extinction crisis, they often contextualize their remarks by referring to the *background rate of extinction*. Using the fossil record as a data source, it is estimated that Earth's species endure, on average, for about a million years before going extinct. Given this background rate, for each million species present on Earth, only one, on average, will go extinct in any given year.

But today Earth is losing species at a rate that is estimated to be somewhere between a hundred and a thousand times faster than this "normal" background level.[10] Indeed, data from the World Conservation Union's Red Book reveal that "27% of the mammals, 40% of amphibians, 34% of reptiles and 14% of bird species" on Earth are now in danger of extinction.[11]

What does this mean for the future? Stuart Pimm, a renowned ecologist at the University of Tennessee, used a combination of field observations, published data, and computer modeling to address this question and concluded that "we might lose between a third and a half of the life on Earth as a consequence of our actions."[12]

But maybe the elevated rates of species extinction that scientists now report and project into the future aren't really anything to be concerned about, because many species seem to do pretty much the same thing. Why do we need twenty different tree species in a forest? Wouldn't one be enough?

Here is a way to think about this question: Imagine that you are traveling overseas and, as you prepare to board your airplane, you notice a guy up on the wing removing rivets. You walk over and yell up, "What the heck are you doing?"

He replies nonchalantly, "Our airline company has discovered that we can sell these rivets for two dollars apiece and we need the money. If we don't pop rivets, we won't be able to continue expanding and offering our customers more services." If you were to encounter such a situation, you would be wise to head back to the terminal, report the rivet popper to the FAA, and book a flight with another carrier—one with a strong commitment to passenger safety.

The point of this example is that all of us, whether we choose to acknowledge it or not, are already riding on a very large carrier, *Spaceship Earth*, and we don't have the option to switch to another craft. The *components* of our "aircraft" are Earth's ecosystems, and our craft's "rivets" are akin to all the species living within Earth's ecosystems.[13]

Removing a few rivets from an airplane wing—that is, driving a few species to extinction—might not compromise the functioning of the airplane (or of an ecosystem). However, it might happen that the sixteenth rivet popped from a wing flap is enough to throw the aircraft into a tailspin, just as removing too many key components from a forest—such as a tree species important in the nitrogen cycle or an insect species important in the pollination of a certain wild fruit species—could compromise the overall well-being of the forest.[14]

In the face of these uncertainties, we would be wise to do everything in our power to heed these words from the American conservationist Aldo Leopold:

> The last word in ignorance is the man who says of an animal or plant: "What good is it?" If the land mechanism as a whole is good, then every part is good, whether we understand it or not. If the biota, in the course of eons, has built something we like but do not understand, then who but a fool would discard seemingly useless parts? To keep every cog and wheel is the first precaution of intelligent tinkering.[15]

Foundation 4.2: Listening to Earth's Ocean Creatures

Early astronauts, looking back to Earth from space, observed something that we land-dwelling humans often overlook—namely, Earth is mostly ocean, mostly blue (see figure 4.1).

Compared to land, Earth's oceans offer an enormous volume of living space for creatures of all sorts. Recent estimates indicate that more than 95 percent of Earth's habitable volume is located in the sea.[16] This enormous

Figure 4.1 An ocean-centric perspective of planet Earth.

volume exists because the ocean has an average depth of two and one-half miles, extending down as much as seven miles in some places. And yet if you are like most people, your encounters with Earth's oceans have probably been limited to beaches, where sea and land touch. Few of us have ventured out into the open sea, much less down to the ocean floor.

As Sylvia Earle (National Geographic Explorer in Residence) points out: "*Green* issues make headlines these days, but many seem unaware that without the 'blue' there could be no green, no life on Earth and, therefore, none of the other things that humans value. Water—the blue—is the key to life."[17]

Life first arose in Earth's seas almost four billion years ago and has continued to diversify and complexify ever since. The nurturing sea gave rise to Earth's first cell, Earth's first plant, Earth's first animal.

Alien World or Cradle of Life?

I'll never forget the first time I went snorkeling. I was in the Caribbean, along a coral reef. Plunging into the water, I was stunned to see big barracudas and groupers nonchalantly cruising around me, but my giddiness was shattered when it occurred to me that I could be a prey item for a rogue shark. This wasn't my habitat; I was out of my element . . . and yet in a way I felt like I belonged because the sea is Earth's ancestral cradle of life. It might even be that each of us still retains some vestigial cellular memory of the sea as our ancestral womb. Consider the experience of eight-year-old Miranda, who sauntered into the ocean one summer day while Mark, her father, looked on:

> Miranda stood in the water up to her waist, just moving back and forth with the waves. Ten minutes passed, and Mark imagined that her eyes were closed. Thirty minutes went by and Miranda was still in the same spot, swaying in the gentle surf. After an hour, Mark found himself swaying with her as he sat and watched from the beach. It was as if she were in a trance. Mark wanted to make sure she was all right. Is this some kind of seizure? Does she have enough sunscreen on? he wondered, but he managed not to intrude.
>
> It was an hour and a half before Miranda came out of the water, absolutely glowing and peaceful. She sat down next to her Dad without a word. After a few minutes, Mark gently asked what she had been doing. "I was the water," she said softly. "The water?" Mark repeated. "Yeah, it was amazing. I was the water. I love it and it loves me." Then, they sat quietly until Miranda hopped up to dig in the sand a few minutes later.[18]

Could what Miranda experienced be available to us all? Could it be part of our birthright to know that our origins are oceanic and that, in a very real sense, the sea is our primordial mother?

FISH FOR ANCESTORS?

You may have heard it said that fish are our ancestors, but really, how could that be? The idea that fish of some primitive species left the water one day and started walking around on land seems utterly fantastical. But, of course, it didn't happen that way. It began gradually, as is often the case with evolution, with fins in certain fish species becoming more like forelimbs and primitive lungs emerging alongside fish gills. These developments would have given some ancient fish species the ability to colonize oxygen-deficient shallow seas and estuaries as well as the capacity to lift themselves out of the water. It is likely that these adventurous fish at first moved back and forth between sea lagoons and land until, eventually, some species shifted their allegiance entirely to land.

Scientists believe the descendants of these land-colonizing fishes gave rise, over vast expanses of time, to such diverse life forms as frogs, dinosaurs, birds, snakes, and mammals, including us. As acclaimed science writer Deborah Cramer reminds us: "Though the fish and the sea seem a world apart and though our lives have substantially diverged from theirs, our origins are oceanic. Their lives gave rise to ours and what they gave us is fundamental to our existence."[19]

"Even if you never have the chance to see or touch the ocean, the ocean touches you with every breath you take, every drop of water you drink, every bite of food you consume . . . Everyone, everywhere, is inextricably connected to and utterly dependent upon the existence of the sea."[20] If you are skeptical, consider: What would your life be like without the oceans? Would you still be able to breathe? Not so well, because a significant amount of the oxygen you breathe is produced by ocean-dwelling photosynthetic organisms! What about water? Would you have enough water to survive? Hardly. Without the vast reservoir of water in the oceans, there would be no rainfall to speak of, no water cycle, and no balanced climate produced by the oceans' ameliorating effects on temperature. Without ocean currents and associated winds, the tropics would be scorched, and the higher latitudes of Earth would be locked into a deep freeze. Indeed, it is the sea that spreads heat from the tropics to more northerly latitudes, creating the conditions for life to flourish.

Exploring the Ocean

Though we have names for Earth's five oceans—Atlantic, Pacific, Indian, Southern, and Arctic—these water bodies are all interconnected by both surface and deepwater currents, creating a single planetary ocean.[21] The global ocean is so vast that only a small fraction (less than 5 percent) has been explored.

Right now, in the spirit of exploration, imagine you are on an expedition boat that has taken you three hundred miles out into the open ocean, beyond the sight of land. The motor has been turned off; the boat is gently rocking. As you gaze into the water, it appears empty of life, but looks can deceive. Your guide takes a drop of seawater and invites you to examine it under a microscope. You are not very good with microscopes and doubt that you will be able to make out anything, but to your astonishment, you see a micro-world teeming with life—hundreds of tiny organisms racing about—in a stunning array of shapes and colors. These are mostly phytoplankton, all busy using sunlight and carbon dioxide to make food for themselves. Although these minuscule organisms make up less than 5 percent, by weight, of Earth's plant life, they account for fully half of Earth's photosynthesis. Translation: The oxygen that phytoplankton produce is part of our every in-breath.[22]

In that same droplet of seawater, you also see zooplankton—tiny animals, mostly crustaceans, grazing on the phytoplankton. Yes, just as cows feed on grassy meadows, zooplankton feed on *meadows* of phytoplankton. Think of them as middlemen, converting teeming masses of phytoplankton into swimming bits of animal protein that serve as food for larger sea creatures.

Until the last century, humans lacked the means to go more than a few dozen feet below the sea's surface, but today it is possible to journey all the way to the ocean floor in a miniature submarine called a *submersible*. Your guide invites you to join her on a trip down to the ocean floor. You squeeze in next to her, filled with a mix of excitement and anxiety.

Fifty feet: A Portuguese man-of-war drifts by and then, suddenly, a dense school of herring appears, racing through the surface waters, feeding on both phytoplankton and zooplankton as they go.

One hundred feet: The sunlight from above catches the water in such a way that you perceive what looks like exceedingly fine *snow* drifting down from above. What you are seeing is actually minuscule bits of dead organic matter slowly settling toward the ocean floor.

Two hundred feet: You are now well beyond the depth that scuba divers can go. Here you see *comb jellies*—semitranslucent creatures with striking bands of iridescent cilia—along with solitary Atlantic sailfish.

Five hundred feet: You are now in the "twilight zone." Looking up, you detect only very faint light; looking below, total darkness, but here, too, there is life. In fact, biologists exploring this zone are frequently discovering new fish species, previously unknown to science.

One thousand feet: You are beginning to enter the ocean depths—Earth's largest habitat. This is the home of the giant squid (*Architeuthis dux*), a creature with eyes as big as Frisbees and a body as long as a school bus.

Looking down into the abyss below, you are startled to see flashes of colored light here and there. These inky depths are home to hundreds of species of fishes and invertebrates, many of whom possess the ability to create light using chemicals from within their bodies.

Why light up? Sometimes the light helps in navigation; sometimes it acts as a warning to back off; sometimes it is a mating signal; and believe it or not, sometimes light can serve as a fishing lure. For example, the anglerfish has a lighted lure extending from the top of its head that draws fish close enough so that the angler can snatch them up.

Ten thousand feet: You have arrived at the seafloor. As you explore this habitat, your attention is drawn to a cluster of six-foot-tall, tube-like structures rooted in the substrate. The "tubes" are topped with red plumes. Your guide explains that these are giant tubeworms—creatures that have neither a mouth nor a digestive system; they survive by using their red plumes to take up oxygen from the seawater along with hydrogen sulfide seeping from hydrothermal vents on the ocean floor. Together these two gases—oxygen and hydrogen sulfide—serve as a primitive food source for the bacteria that comprise up to half the tubeworm's body weight. These bacteria, through a process known as chemosynthesis, oxidize the hydrogen sulfide and create energy (organic food molecules) that tubeworms subsist on.[23]

Ocean sediments: Below the ocean floor are sedimentary layers extending down thousands of feet. Scientists, drilling into these sediments, find that they are loaded with microbes, some distinct in their lineage from all other known microbe groupings. The metabolism of these minuscule creatures is so slow that they might only reproduce (by dividing in half) once every thousand years.[24]

As you take one last look around before ascending to the surface, you are surprised to see an empty Coke can, partially buried by sediment. Though very few humans will ever get to go to the bottom of the ocean, our trash is already there.

Checking the Health of the Ocean

Given the immensity of Earth's oceans it seemed impossible, at least until recently, that humans could harm this vast ecosystem, but we see now that we were wrong. Over the past hundred years, humankind has been engaged in a steady assault on Earth's oceans.

Ocean Fisheries under Attack

Each year, worldwide, approximately 160 billion pounds of fish are harvested from the world's oceans. This amounts to roughly 40 pounds of fish harvested for each person alive today.[25] Since the 1950s, the combined harvest of wild and aquaculture-raised ocean fish has increased by roughly five-fold.[26] Because of this precipitous increase, scientists now estimate that 90 percent of the ocean's wild fishes are in danger of being exploited beyond what is sustainable.[27] Some of the ocean's large-bodied fish species—for example, Atlantic halibut, monkfish, and bluefin tuna, along with many species of sharks—are already suffering population declines as a result of highly aggressive industrial fishing technologies.

Worldwide, tens of thousands of large fishing boats scour the ocean each day. Many of these vessels use long lines that trail behind for dozens of miles, with as many as three thousand baited hooks distributed along each line. Another industrial technique involves trawlers that drag enormous nets—as wide as a football field and taller than a three-story building—along the sea bottom. These heavy-duty nets act like bulldozers, scraping along the ocean floor, scooping up everything in their path. On a yearly basis, an area of seafloor equivalent to the size of the continental United States is decimated by trawlers. If this plundering were happening on land, there would be an outcry against it, but because it's invisible, it continues unabated.[28]

Both long lines and trawlers inadvertently capture many sea creatures that are not of commercial interest.[29] These days, for every pound of fish that goes to market, ten or more pounds are thrown away as *bycatch*. It's even worse in the case of shrimp; for every bushel of shrimp harvested by trawlers, roughly a hundred bushels of sea life are scraped from the seafloor, only to be discarded.

As stocks of large-bodied commercial species are exhausted, smaller fish like anchovy, shad, and mullet—which large fish would normally be feeding on—have multiplied and become a new target for commercial fishing operations. If/when populations of these smaller fish species become depleted, it is possible that the whole ocean ecosystem could begin to fall into disarray. What might that look like? In the words of biologist Tierney Thys, "With fewer fish . . . animals such as *jellies* can bloom in great numbers and gain a stronghold on the sea."[30] Thys's words evoke the airplane-rivet-popping metaphor from earlier in this chapter. By removing the ocean's *rivets*—that is, by overexploiting certain fish species—humans may inadvertently be disassembling the oceans, making Mother Earth less healthy, less whole, less home.

Ocean Dead Zones

In the early 1970s, biologists surveying the waters of the Gulf of Mexico, where the Mississippi River flows into the Atlantic Ocean, were surprised to discover that the water at the bottom of this estuary had a milky appearance and was mostly devoid of sea creatures. When the researchers analyzed the bottom waters, they found high concentrations of nitrogen and phosphorus, but almost no oxygen.

Put yourself in the place of the researchers attempting to solve this mystery. Why were the bottom waters enriched in nitrogen and phosphorus? And why was the area almost devoid of oxygen? As you ponder these questions, recall what you learned in chapter 3 about how disturbances, like deforestation, often interfere with natural cycles, causing nutrients (e.g., nitrogen, potassium, phosphorus) to leak from the land into streams and rivers. Now picture the massive amounts of water—more than a million gallons every second—flowing out of the Mississippi and into the Gulf of Mexico. This outflow originates in the form of rain falling on the thirty-one farm states that constitute the enormous Mississippi River Basin.

Before the advent of agriculture, the waters of the Mississippi ran clear and clean, but now that much of the Mississippi Basin has been cleared for farming, enormous quantities of eroded soil and fertilizers make their way into the Mississippi year after year. When these nutrients reach the Gulf, they act as *fertilizer*, provoking explosions in algal populations. Later, when the masses of algae die and settle to the bottom of the Gulf, they serve as food for bottom-dwelling populations of microbes (decomposers). Inundated with food, the microbe populations explode, eventually exhausting the oxygen supply in the bottom waters. This in turn creates a vast dead zone where bottom-dwelling sea creatures, including oysters, mussels, and clams, are no longer able to survive.[31]

The dead zone in the Gulf of Mexico has been expanding since the 1970s and now covers roughly six thousand square miles.[32] This phenomenon is not exclusive to the Gulf. Dead zones have been identified virtually everywhere on Earth where major rivers flow into the ocean.[33]

There is more to this story. Based on a major study coordinated by UNESCO's Intergovernmental Oceanographic Commission, there are now large expanses of open ocean with oxygen-starved waters. Though fish can still survive in these oxygen-depleted waters, their ability to grow and reproduce is compromised.[34]

Plastic Oceans

If you have ever gone beachcombing, chances are you have encountered bits of trash among the shells and driftwood. That is not surprising insofar as we, humans, are sometimes careless with our trash. What would not be expected—at least I never imagined it—is that enormous patches of human-generated trash—consisting mostly of buoyant plastics—now exist far out in the open ocean.

An estimated eight million tons of plastic waste finds its way to the world's oceans each year, accumulating in enormous vortices created by oceanic circulation patterns.[35] The most famous patch—dubbed the Great Pacific Garbage Patch—is located halfway between California and Hawaii and extends over an area of 600,000 square miles (twice the size of Texas).

Some years back Charles Moore pulled a fine mesh net through the Great Pacific Garbage Patch and then sorted through the contents, comparing the weight of living organisms (mostly plankton) to plastic debris. The result? Life lost 6:1—that is, for every pound of sea creatures there were six pounds of trash, almost all of it plastics.[36] The Great Pacific Garbage Patch is one of five such ocean zones that concentrate plastics. Combined, these trash-collecting gyres represent 40 percent of the sea, with plastics making up 90 percent of their contents.[37]

As recently as the 1950s, plastic was rarely seen in the United States, but nowadays industries turn out more than three hundred million tons of the stuff each year, with half of this output trashed after a single use. In per capita terms, those living in the United States throw away, on average, two hundred pounds of plastics each year.[38] The way things are going, by the middle of this century there will be more plastic, by weight, in the ocean than fish.[39]

All of this plastic debris won't disappear anytime soon. Indeed, except for the minuscule amount that's been incinerated, every molecule of plastic ever created is still around.

This plastic waste has harmful effects on sea life, in part because many plastic fragments look like food to sea animals and, once ingested, can create obstructions that cause choking and intestinal clogging, leading to death.[40] Each year an estimated 100,000 marine mammals—for example, dolphins, seals, porpoises, whales—die because of plastic pollution.[41]

The actual chemical makeup of plastics is also a concern because plastics are imbued with chemicals (e.g., dyes, flame retardants, softeners), and none are intended to be eaten. Moreover, the open molecular structure of plastic fragments in the ocean makes them *sponges* for the absorption and concentration of common pollutants like PCBs (polychlorinated biphenyl) and PAHs (polycylic aromatic hydrocarbons.)[42] One need only consider eating a plate of oysters laced with minuscule, toxin-laden plastic particles to imagine how a plastics-infused ocean could come home to roost in our very own bodies, especially given that 99 percent of the plastic waste in the ocean is invisible to the human eye![43] This is just one more reminder of how our lives, for better or worse, are inextricably bound to the life of the sea.

Stepping Back to *Feel* the Big Picture

The depth of our feeling life measures the depth of our life force, and if we judge, contain, or repress our feelings we repress our life force.

—Ingrid Bacci[44]

Having just read about how plastics are contaminating the world's oceans, can you engage your capacity for *umwelt* by imagining what a whale might think and feel when she encounters—as surely she must—the Great Pacific Garbage Patch? Similarly, do you have the ability to put yourself in the place of a migrating songbird who, upon returning home in the spring, discovers that the forest that has been her nesting place for years has disappeared, been demolished? Your response to these questions will tell you if you have the bigness of heart and mind to extend empathy to the feathered and the finned ones with whom we share Planet Earth.

As we each begin to open to the battered condition of our home, Earth, some common responses are:

- Why didn't anyone ever tell me about this stuff before?
- I want to scream when I think about what humans are doing to our Mother, Earth.
- I care a lot about Earth and what we are doing to her, but no one else seems to care.

Might it be, though, that hundreds of millions of people (perhaps billions) around the world share these concerns and sensitivities, but most remain silent, reluctant to voice their feelings publicly?

Eco-psychologist and activist Joanna Macy believes that fear is often at the root of our silence in these matters. For example, we fear that if we speak of our pain concerning the plight of Earth, others will judge us as weird, negative, and/or overly emotional and avoid our company.[45] To this Macy contends: "Pain is the price of consciousness in a threatened and suffering world. . . . Feeling pain is not only natural, it is an absolutely necessary component of our collective healing. As in all organisms, pain has a purpose: It is a warning signal, designed to trigger remedial actions."[46]

We can begin to free ourselves from the bondage of fear and despair by giving ourselves permission to feel. What arises in you when you read about dead zones in the ocean, plastic-infested seas, fragmented forests, dying

TO FEEL IS OUR BIRTHRIGHT

As a male growing up in the United States, I was taught to regard my feelings as something to keep under cover or, at the very least, under control. In school, when I fidgeted in my seat, I was told to sit still. When I fell and skinned my knee in the schoolyard, I was told that big boys don't cry. If I expressed anger, I was told to get myself under control. It's as if there was a sign at the door to my elementary school saying, "Check your emotions at the door!"

Looking back, I now understand what was going on. When I was in touch with my emotions, I felt alive and my actions arose from within me, but in school my job as a student was to sit still, face forward, and follow instructions. To be spontaneous, to express feelings of excitement, joy, disappointment, or sadness, was pretty much off limits.[47]

birds? Specifically, what happens in your body—in your jaw, your neck, your shoulders, your heart? What happens with your breathing? What about your thoughts? Is there judgment, resistance, anger, fear? Do you want to run away? Do you feel helpless? Numb? Indifferent?

Sometimes, as a way to help students express their capped-over feelings regarding the battered condition of Earth, I invite them to gather with me in a circle. The circle contains four objects: (1) a cold rock to remind us of what our hearts are like when we are afraid; (2) a fierce stick to channel and shake loose our anger; (3) a dry, shriveled leaf to call forth our sadness; and (4) an empty bowl to connect us with feelings of emptiness and despair. Imagine yourself, now, sitting in that circle. If the destruction of Earth creates fear in you, you could pick up the hard, cold rock and, holding it, speak of your fear. If it is anger that you feel, you could grab the stick and, holding it fiercely in both hands, give voice to your anger. If it's sadness that envelops you, you could hold the dead leaf and speak of your heartbreak. Finally, if you are gripped with emptiness and despair, you could cradle the empty bowl, putting words to whatever arises.[48]

The reason for this circle work is to bring awareness to the capped pain that we all carry to varying degrees. We carry this pain because we are, inescapably, an expression of the living—and in these times, wounded—body of Earth. As psychologist Ingrid Bacci points out, "Our world is in a mess right now because too many of us have lost touch with the art of caring. We need to learn how to care, and there is only one way to begin doing that—to practice experiencing our feelings."[49]

Wrap-Up: Expanding Ecological Consciousness

> No other quality is so urgently needed today [as listening]: Millions cry out to be heard, to be listened to, to be allowed to be. . . . Planet Earth [herself] cries out amid [her] pain of pollution, exploitation and desecration.
>
> —Diarmuid O'Murchu[50]

Environmental scientists have been listening to the creatures of the land, of the sky, and of the sea, and the messages they have been receiving reveal that we have been abusing our home, our Mother . . . but this need not be. We are not mindless automatons; we are bearers of reflective consciousness. Our very name, *Homo sapiens sapiens*—that is, the ones who know that we know—bears testimony to our capacity for awareness, connection, and awakening.

We are being challenged to see ourselves and the world with fresh eyes. Instead of being mired in *speciesism*—the trap that has us believing that Earth is simply here for us to use as we see fit—we can choose to adopt an *ecocentric* worldview by valuing and respecting all of Earth's species and ecosystems to the same degree that we value and respect each other and ourselves.[51]

An ecocentric orientation calls us to embrace the Noah principle, which holds that a species' long-standing existence on Earth carries with it unimpeachable rights to continued existence and respect. Although humans have not yet adopted the Noah principle, we have taken small steps in this direction. For example, during its history the United States, not without struggle, has extended rights and acknowledged moral obligations to previously disenfranchised segments of society, including women, children, people of color, and those who identify as LGBTQ+. Likewise, through legislation such as the Endangered Species Act, we have begun to consider our moral obligations to the plant and animal species that populate Planet Earth.

Ideally, this progression toward an ever-wider and more inclusive sense of kinship with the world unfolds over the course of one's lifetime (see figure 4.2).[52] As young children, our sphere of awareness and concern is mostly

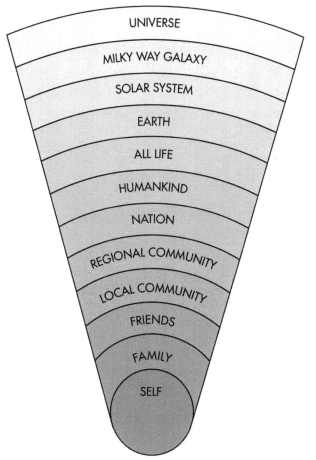

Figure 4.2 The expansion of human awareness from childhood through late adulthood.

restricted to our immediate families. In grade school, our awareness expands to schoolmates; in high school, our purview of concern begins to extend outward to our community; and by adulthood, if all goes well, we begin to experience ourselves as members/citizens of a nation. With still further expansion of our consciousness, we come to see ourselves as integral members of Earth's community of life. When this happens for humankind as a whole, we will be dwelling in ecocentric consciousness. Getting there won't be easy, but it is possible. As author Barbara Kingsolver once said to her daughter: "We can do hard things."

Applications and Practices: Listening

> Learning how to communicate with animals is just like learning any other language. The more you practice, the better you become.
>
> —Karen Anderson[53]

Listening to the Wild World Beyond Your Doorstep

If it's intimacy with nature that you seek, you don't have to go far. The next time you step outside, leave the pavement behind and wander, going where your feet take you. Then, when the time feels right, lower yourself to the ground and close your eyes, gently bringing your attention to your breath. With each exhalation, release any tension that you might be holding in your shoulders, your neck, your back, and so forth. Then open your eyes and, with a soft gaze, slowly take in your surroundings, until your attention is drawn to something. It might be a leaf on the ground, an ant wending her way through tall grass, a roving bee, or a patch of moss. Notice how the subject of your attention is simply himself/herself—utterly real, authentic, with no pretense.

Open to the possibility that you are there to learn something from this *Other*. There is no rush. Be curious . . . open; Soften . . . breathe . . . and then, as a matter of politeness, take the bold step of introducing yourself, saying, out loud: *I am* (your name), followed by *I am the one who* (voicing whatever arises), and *I am here to* (again, voicing whatever comes up). Then, settle in with this other, allowing your senses and innate curiosity to carry you into relationship. As you relax, consider the connections—the web strings—between the two of you. The idea is to take in and receive this other with your whole being—eyes, ears, skin, heart, intuition, belly. You are not waiting for this other to necessarily *tell* you anything; rather, you are simply present, curious, and available for relationship, witnessing the individuality of this other, open to whatever emerges.

CURIOSITY AND HUMILITY AS GATEWAYS TO CONNECTION WITH THE MORE-THAN-HUMAN WORLD

Primatologist Barbara Smuts learned to see animals as individuals when she lived with a baboon troop in the wilds of Kenya. After several months of meticulous observations, she was able to identify all one hundred–plus baboons in the troop as distinct individuals, based not just on their physical appearance, but also on unique behavioral characteristics and mannerisms.

A short time after she achieved this level of familiarity, one of the dominant males in the troop looked in her direction and flattened his ears against his skull while simultaneously raising his eyebrows. Smuts knew that this was a signal used by male baboons to call females to them. She turned around to identify the female baboon that the male was signaling, but to her astonishment, there wasn't a baboon behind her. The male was actually signaling to Smuts. She proceeded to act, in a baboon-appropriate way, to decline the male's advance.

In retrospect, what Smuts found most remarkable from her time living with baboons was that they didn't acknowledge her as a person until they had had the experience of being seen, as unique individuals, by her. This insight underscores that curiosity and humility are gateways to relationship with the more-than-human world.[54]

Listening to the Human Other

These days, when suffering is everywhere—in the wars that divide nations, in the exploitation that defiles ecosystems, in the psychic depression that shrivels hearts—we have no bigger challenge than to learn how to listen to one another.

Listening can be especially challenging when we are hearing things with which we disagree. For example, imagine the following scenario: You are sitting in a coffee shop when the guy at the next table throws down his newspaper and exclaims, "Those damn Enviro-tree-huggers are clueless! They are trying to stop everyone from making a living."

You look over and catch his eye, signaling a willingness to listen. Being fully present with him is not going to be easy because your sympathies tend to be more with the *enviro-tree-huggers* that he is condemning. But instead of getting hooked by the man's words, you resolve to place your attention on the feelings that lie below his upset. So it is that you are able to nod empathetically and observe, "Seems like you are really upset."

He shoots back, "I am fed up with their protests. They are a bunch of spoiled brats and they don't know jack."

"They really piss you off," you affirm.

"Damn right, I'm pissed. My brother and uncle are both loggers, and their jobs are on the line if this shit keeps up."

You pause; and once again, focusing on what you hear as this man's feelings, you say, "You're worried about how these protests could interfere with the well-being of your people."

"Yeah, you know it's not easy being a logger these days," he responds.

"And you're worried about your brother and uncle," you add.

"Well, yeah. They're family!"

As you listen without judgment, you are able to depersonalize the man's initial complaint about the "enviro-tree-huggers," coming to appreciate that his anger toward the protesters is born of his worry and concern for his people.

Once we understand that our judgments of others (in this case, the logger judging the environmentalist) are rooted in our own unmet needs, it becomes easier to bridge differences. For example, continuing the conversation, you could move from the logger's *feelings* to his *needs* by saying, "Sounds like it is important for you to be assured that your brother and uncle aren't going to be out on the street without jobs."

To this the man might respond: "Yes, they both have young kids. Without a job they'd be screwed."

As you continue to listen, it becomes clear that this man doesn't harbor any ill will toward environmentalists per se. In fact, he confesses that he likes to fish and that all the runoff and erosion associated with forest clear-cutting is messing up his favorite trout streams.

This scenario validates the old adage: If you hate someone or condemn someone, you just haven't heard their story yet. Listening without judgment creates enormous space for understanding and peacemaking.

If you are game, you could test this by identifying someone in your life who is causing you stress—it could be a friend, a parent, a brother or sister, a boss. Write down all the things about this person that you find upsetting. Include everything about the person that angers, disappoints, and confuses you. Don't hold back. Finish by describing how you want this person to change.

When you are done, read over what you have written and consider how some of your judgments about this person might not be true. After all, judgments are beliefs, and beliefs are really just opinions. So, see if you can soften a bit. In a similar vein, open yourself to the possibility that some of your judgments about this person might actually apply as much to you as to him/her. For example, if you are upset with a housemate because s/he ignores you, see if you can identify instances when you have ignored this housemate. Again, the idea is to see if you can soften and open a bit. This kind of opening is not easy, insofar as our egos are invested in blaming and judging and ensuring that we are the ones who are right.

Now for the really challenging part. Spend an hour with this person who is causing you stress with the sole intention of understanding her/him—free of judgment, free of your need to be right. In carrying out this exercise, you might discover that it's hard to be curious about someone while also standing in judgment of them.[55]

Questions for Reflection

- What, if anything, would you do if you happened to see a cat stealthily creeping up on a migratory songbird? Why?
- Where were your favorite nature hangouts as a child? What has happened to those special places? Do you know? Do you care?
- What feelings come up for you when you think about the Great Pacific Garbage Patch? Where in your body do you experience those feelings?
- In what ways has your voice been silenced by fear (à la Joanna Macy), and what, if anything, might you do about this?
- Have you ever referred to Earth as your *Mother*? And in this same vein: What would it mean for you to hear, within yourself, the sounds of the Earth's crying?
- What are your most common judgments of others, and what personal unmet needs might lie below these judgments?

5

Courage

Facing Up to the Unraveling of the Biosphere

> It is obvious that something is not working properly. How else can we explain our mismanagement of resources and of our own population, the pollution and the destruction of our environment, or the mass murder of our own species? We cannot see the bigger picture and how we fit into it. We are no longer in our bodies; we are not in our right minds.
>
> —Wes Nisker[1]

The subtitle of this chapter—"Facing up to the Unraveling of the Biosphere"—has an unnerving quality. Could it really be true that the Earth that we inhabit today is different from the Earth that existed, say, in 1700? I don't mean *different* because we now have all kinds of nifty technologies; I mean *different* from the perspective of Earth—*different* biologically, functionally . . . less whole, less healthy, less stable?

Here's an analogy. When you become sick, if you are like most people, you behave differently. For example, you don't move the same way, don't eat the same way, don't act the same way; you are off-kilter. After a time, though, your body begins to heal itself and eventually, with any luck, you are back to normal. But imagine that you didn't recover, that instead, you became sicker and sicker. This aptly describes the trajectory that Earth is now on.

It appears that we are now *perilously close* to pushing Earth beyond her capacity for self-regulation—that is, beyond her ability to self-heal on a time scale relevant to human flourishing. Many scientists fear that we may have already gone too far; that is, that we now live on a feverishly sick planet that will become increasingly inhospitable to humans, including you and me, in the years ahead.

The challenge in this chapter is to summon the courage to look, full-on, at our human prospects, in light of such global phenomena as climate change; eroding soils; dwindling aquifers; and the growing contamination of the air we breathe, the water we drink, and the food we eat. None of this is easy to face, but if we are to survive—and our survival is by no means certain—it is time to shake off the shackles of complacency and denial, time to wake up!

Foundation 5.1: Climate Out of Control?

Of all the worrisome environmental issues that we face, none is more troublesome than climate change. Earth's climate is changing; this can no longer be denied. Much of this change is being triggered by the steady increase in carbon dioxide in Earth's atmosphere, as evidenced in figure 5.1.[2] This graph reveals a steady upward trend in atmospheric carbon dioxide concentrations in the atmosphere since the late 1950s.

A fascinating aspect of this graph is the see-saw—rise and dip—in carbon dioxide concentrations each year. These steady oscillations mirror, in a sense, Earth's *breathing*. To understand what is happening, imagine measuring the annual fluctuations in atmospheric carbon dioxide in a single patch of forest in the temperate zone, for example, in Pennsylvania. Begin by picturing the forest in winter. The leafless trees have no means of gaining energy through photosynthesis and so, to stay alive, they must break down sugars stored in their stems and roots. In the process, they release carbon dioxide (CO_2) to the atmosphere. All the animal life in the forest patch will also be respiring—that is, breathing out CO_2. Hence, there will be a net increase in the carbon dioxide level of the atmosphere during the winter. Now, fast-forward to spring when all the trees flush out new leaves. Since

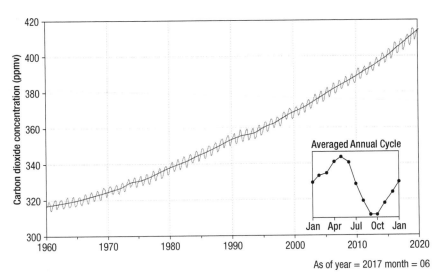

Figure 5.1.   Increases in Earth's atmospheric carbon dioxide concentrations since 1958.

carbon dioxide is necessary for photosynthesis, the new leaves will begin to take in carbon dioxide from the surrounding atmosphere. Of course, all the trees and animals will continue to release some carbon dioxide to the atmosphere via respiration, but the carbon dioxide uptake by the photosynthesizing trees will be much greater than what's released through respiration. Thus, there will be a net decrease in CO_2 in the atmosphere of the forest patch during the growing season. This same dynamic is happening on a global scale. Because there is more land and more trees in the Northern Hemisphere, Earth *breathes in*—that is, removes carbon dioxide from the atmosphere—during the Northern Hemisphere's spring and summer (creating the dips in figure 5.1), and *breathes out* carbon dioxide during the Northern Hemisphere's fall and winter (creating the rises).[3]

There is another point to be garnered from this graph—namely that carbon has its own cycle. Just like calcium and other elements (see chapter 3), carbon moves among various storage places or pools. The main pools for carbon are Earth's forests, soils, oceans, and atmosphere. For almost all of human history these carbon pools have been fairly stable, but starting about two hundred years ago, humans began to tap into enormous new pools of carbon in the form of oil, coal, and natural gas. Burning these fossil fuels for energy has resulted in the utter transformation of human life and of the environment. Here's why:

> One barrel of oil yields as much energy as twenty-five thousand hours of human manual labor—more than a decade of human labor per barrel. The average American uses twenty-five barrels each year, which is like finding three hundred years of free labor annually. And that's just the oil; there's coal and gas too . . . [The burning of fossil fuels is why] we're prosperous, why our economies have grown. It's also, of course, why we have global warming and acid oceans; in essence we've spent two hundred years digging up all that ancient carbon, combining it with oxygen for a moment to explode the pistons that take us to the drive-through, and then releasing it into the atmosphere, where it accumulates as carbon dioxide. . . . All day every day we burn coal and gas and oil, from the second we make the coffee till the second we turn out the lights. . . . If an alien landed in the United States on some voyage of exploration, he might well report back to headquarters that we were bipedal devices for combusting fossil fuel.[4]

Learning to extract and burn fossil fuels has been a tremendous boon to humankind. However, nowadays our relentless combustion of these fuels is causing Earth's climate to warm, with potentially disastrous consequences. This should come as no surprise. After all, we've seen time and again in this book that everything is connected—how it is impossible to do even one thing without creating ripple effects.

MAKING THE INVISIBLE VISIBLE

Fossil fuels are integrated so seamlessly into our lives that chances are you have never bothered to consider what your life would be like without them. So take a moment, now, to do this by considering the energy that is contained in just one gallon of gasoline. If you were to put that gallon in a typical car, it would move the car twenty-five miles. But what if, instead, you had to rely on the energy within your body to move (push) that car those twenty-five miles? How long do you suppose that would take you? A day? A month? A year? Granted, there are a lot of variables to consider (e.g., car size, terrain, road surface, body strength), but all things being equal, the energy required would be equivalent to about three hundred hours of human labor (think seven work weeks). By contrast, a gallon of gasoline would propel you those twenty-five miles in just a half hour. Of course, if you pushed, you'd save three dollars in gas money, but not really, because the monetary returns on your labor, if you were pushing, would amount to only a penny per hour (300 hours in labor/$3.00 for gallon of gas = 1 cent/hour).[5] No wonder fossil fuels have transformed the world.

What Exactly Is Global Warming?

The science of global warming is grounded in four simple facts: (1) energy enters Earth as sunlight and leaves in the form of heat (i.e., infrared radiation); (2) for Earth's temperature to remain constant, the energy reaching Earth, as sunlight, must approximate the energy that leaves Earth as heat; (3) when we burn fossil fuels, the carbon atoms in these fuels combine with oxygen in the air to create the gas carbon dioxide; and (4) the molecular structure of carbon dioxide is such that it traps heat that otherwise would escape to space, and it is this heat trapping that is upsetting the energy balance of our planet.

Some trapping of infrared radiation is actually a good thing. Without it, Earth would be brutally cold, with the ocean frozen from top to bottom. At the other extreme, if none of the solar radiation entering Earth's atmosphere was able to escape as heat, Earth would quickly warm up to unbearable temperatures. Maintaining the right balance is important. Our voracious consumption of fossil fuels has upset this balance. Indeed, "by most accounts, we've used more energy and resources during the last 35 years than in all of human history that came before."[6]

What Does Climate Research Reveal?

If one of your ancestors had climbed to the top of a remote mountain in the United States in 1850 and collected a bottle of air, and if you were to go to

HOW COULD THIS BE TRUE?

It seems like an outrageous proposition to suggest that we, mere humans, could actually change the climate of something as grand as Planet Earth. But Earth's atmosphere doesn't extend endlessly out into space; it is only a wafer-thin protective layer. How thin? Imagine Earth scaled down to the size of a basketball. Dip that ball in water and pull it out. That gossamer-thin sheen of water adhering to the ball represents, to scale, the extent of Earth's atmosphere. Human activities can and do affect this narrow, life-sustaining band.

that same mountain today and collect a new sample, the two samples would be different. Yes, "the air around us, even where it is clean and smells like spring and is filled with birds, is different, significantly changed."[7]

The big difference, as illustrated in figure 5.1, has to do with rising carbon dioxide concentrations. In 1850, Earth's atmosphere contained about 280 parts of carbon dioxide per million;[8] this was before humans started burning fossil fuels in earnest. Since that time, as emissions from fossil fuel burning have steadily increased, so has global warming.

Since scientists first began recording Earth's temperature 140 years ago (1880), the twenty warmest years have all been since 1995.[9] That's a sentence that you might want to read again.

Not only has the amount of carbon dioxide in the atmosphere increased steadily in recent decades, but the *rate* of increase has also been accelerating. For example, in the ten-year period 1992 to 2001, the average rise in atmospheric CO_2 concentration was 1.6 ppm/year; in the following decade (2002–2011) the average rise increased to 2.1 ppm/year; and at present, the annual rise is in the 2.5 ppm range.[10]

At the time of this writing, the CO_2 concentration in Earth's atmosphere has topped 415 ppm, an almost 50 percent rise since 1850, and higher than at any point in the last fifteen million years.[11] If current trends continue, atmospheric carbon dioxide concentrations will top 500 ppm before 2050. If this happens, most climate scientists concur that all human life-support systems on Earth will be in significant disarray, challenging humankind to survive in a more chaotic world than we have ever known.

The Intergovernmental Panel on Climate Change (IPCC), established by the United Nations in 1988, is charged with integrating research findings from more than two thousand climate scientists representing more than 120 nations. The most recent IPCC report confirms that (1) the scientific evidence for climate warming is "unequivocal"—meaning virtually certain; and (2) immediate action, on a massive scale, must be taken NOW to avoid irreversible changes to Earth's climate system.[12]

**DOES A WARM SPELL IN JANUARY
PROVE GLOBAL WARMING IS REAL?**

Weather changes hour by hour, day to day, and in the course of any given year, there will always be days that are unseasonably warm or cold. Translation: A spell of warm weather in winter doesn't prove global warming is occurring any more than a period of unseasonably cool weather in summer disproves the reality of global warming. Weather, after all, is a short-term thing (think hours, days, months); climate, on the other hand, is a long-term phenomenon (think decades, centuries, millennia). There is nothing new about weather change—that's a given; what's new for those of us alive today is climate change!

What Does Climate Change Look Like Around the Globe?

You might have been thinking that global warming just means warmer days, offering more time at the beach. Think again; in its most fundamental expression, global warming means that the amount of energy in Earth's climate system is steadily increasing. By analogy, it's like turning up the volume, bit by bit, on a stereo system, until you reach a point beyond what the speakers can handle. When this happens, the result is ever-increasing distortion (static) and destabilization. The same sort of disruption is happening within Earth's climate system. We have been ramping up the volume (think heat energy) in Earth's atmosphere over the past 150 years, and this change is leading to planet-wide destabilization. So it is that scientists now advise us to prepare for more climate upheaval, not more sunbathing at the beach.[13] Climate chaos is now a reality not just to scientists but to all of us.

For many around the world, heat waves are the most palpable manifestation of warming. Indeed, just since 2000, we have experienced nine of the ten deadliest heat waves in human history.[14] Yes, there are limits to the amount of heat that you and I can withstand. A study by Australian and American researchers determined that when temperatures of 95 degrees F (or more) are coupled with humidity levels above 95 percent, "the body can't cool itself and humans can only survive for a few hours." The researchers went on to determine that 1.5 billion humans (a fifth of humanity) now live in regions where such temperature conditions are increasingly common, creating increased risk of heat stroke, as well as famine and mass migration.[15]

Consider the Arctic Summer Ice

Since the 1970s, scientists have observed that the Arctic is warming just as fast as, or even faster than, the rest of the planet.[16] Visual evidence of this warming came in 2007 with the news that large zones of the Arctic summer ice had melted. Since then, the area of Arctic summer ice cover has declined by more than 40 percent and may disappear entirely by midcentury, given current trends.[17] At present, the Arctic ice-free zone in summer covers about 1.3 million square miles, an area more than one-third the size of the continental United States.

Previously, when ice remained intact during the Arctic summer, almost all of the solar radiation hitting the white surface of the ice was reflected back into space; now, with the steep decline in summer ice, solar energy is being absorbed by the dark surface of the exposed ocean.[18] The result is both warmer water and greater transfer of seawater to the atmosphere via evaporation, both of which contribute to climate destabilization.[19]

Consider the Tundra

Travel south from the Arctic Circle and you will enter vast swaths of treeless tundra vegetated by ground-hugging lichens, mosses, and shrubs. Below the tundra's thin surface soil is a zone of frozen ground (*permafrost*) that is sometimes hundreds or even thousands of meters deep. The permafrost is composed of soil, rock, and carbon-rich organic matter that has accumulated over tens of millions of years. This organic material, referred to as "peat," is in the early stages of coal formation. Until recently, these peat deposits—estimated to contain fifteen hundred billion tons of carbon (roughly twice the carbon now in the atmosphere)—were cloistered away in what amounted to a *cold-storage locker*. But now, as the Arctic heats up, the upper layers of permafrost are beginning to warm and dry out in many places.[20] If thawing of the permafrost continues, microbes will gain access to huge stores of carbon-rich organic matter, resulting in the potential release of enormous quantities of CO_2 to the atmosphere. This is especially worrisome given that permafrost covers one-fifth of the Northern Hemisphere.[21]

Warming and drying of the tundra is also contributing to wildfires such as the ones that incinerated thirteen thousand square miles of tundra in Alaska and Siberia in 2019.[22] Conditions for such fires have not existed for the past ten thousand years (i.e., since the end of the last ice age), but climate models now predict a significant increase in wildfires in the Arctic tundra in the years to come.[23] Whether it is through permafrost melting or through fire, the threat is the same: an enormous carbon pool may soon be finding its way into the atmosphere as CO_2 gas.

The conversion of peat to greenhouse gases is also occurring in the subarctic region of Western Siberia. There, an expanse of once-frozen bog land (the size of France and Germany combined) contains an estimated seventy billion tons of methane, a greenhouse gas with twenty times the heat-trapping capacity of CO_2. This methane is now bubbling out into the atmosphere.[24]

Consider Sea-Level Rise

In recent years, global sea level has been slowly rising. For example, the rise for the twenty-five-year period 1993–2018 was three inches, double the rate for the twentieth century as a whole.[25] A few inches or even a foot rise in sea level doesn't seem like much to worry about, but consider that approximately 40 percent of the U.S. population lives in coastal regions that are prone to hurricanes, storm surges, and calamitous flooding, all of which can wreak havoc on human infrastructure, including transportation systems, water and sewage supply networks, power plants, and more.

The recorded sea-level rise up to this point has been due to the thermal expansion of seawater as it warms, but there is another much more worrisome potential contributor—the addition of water to the ocean from melting ice sheets and glaciers. Scientists caution that sea level rise could turn catastrophic if the vast ice sheets and enormous glaciers of Antarctica and Greenland begin to unleash their meltwater. In fact, the melting of just one glacier could have calamitous consequences. For example, NASA scientists have discovered that the Thwaites glacier (roughly the size of Florida), located on the West Antarctic ice sheet, has developed a massive subterranean hole—an ominous sign that this glacier may be melting much faster than expected. According to some current climate models, if the entirety of the Thwaites glacier were to melt over the next century, sea level could rise by as much as ten feet. As meteorologist Eric Holthaus warns, "This could happen in the lifetimes of people alive today, flooding every coastal city on Earth and potentially grinding civilization to a halt."[26] Is this likely to happen? The jury is still out.[27]

Consider the World's Forests

Forests play an important role, acting as a buffer against global climate change because, as they grow, they take carbon dioxide out of the atmosphere, storing it in their roots, trunks, and branches. However, today forests, instead of serving as depositories or *sinks* for atmospheric carbon, sometimes act as a *source*, releasing carbon to the atmosphere, aggravating rather than

---

A WARMER WORLD IS A WETTER WORLD

Since the 1980s, global rainfall has been increasing by about 1.5 percent each decade. This increase has occurred because warmer air holds more moisture.[28] Meanwhile, the spatial distribution of rainfall is changing because of the destabilizing effects of greater energy in Earth's climate system. For example, some places, like North Africa, are now experiencing more frequent and severe droughts, while other regions, like Southeast Asia, are being confronted with more frequent and severe storms and floods.

ameliorating climate change.[29] For example, throughout the western United States and north into British Columbia, there are now millions of acres of forests composed of dead and dying ponderosa and lodgepole pines. The principal culprit in this massive forest dieback is an insect known as the *pine bark beetle*, which makes its home in the sapwood of pines, killing them by clogging their vessels.[30] Before climate warming became a problem, pine beetles suffered severe diebacks in winter, and this alone was enough to hold their populations in check, but now, with climate warming, pine beetles are spreading far and wide.

Meanwhile, all the dead pines, both standing and collapsed, represent an enormous stock of carbon-rich fuel. Add a spark, and you can easily get a forest fire. Global warming is contributing those sparks because, as the planet heats up, the frequency of lightning strikes—think match strikes—has been increasing. Between six and ten million acres—mostly in the West—are now impacted by wildfires each year, up from three or four million acres/year in the 1990s.[31]

Just since 2000, more than a dozen states have recorded the worst wild-fires in their histories.[32] "These blazes make up a new category of fire, exhibiting behaviors rarely seen by foresters or firefighters. . . . The flames create their own weather systems, spinning tornadoes of fire into the air, filling the sky with pyro-cumulus clouds that blast the ground with lightning to start new fires, and driving back firefighting aircraft with their winds."[33]

Contributions to climate warming resulting from forest loss and fire are also at play in the tropics, which contain half of Earth's forest biomass.[34] The situation in the Amazon basin is particularly worrisome because this extraordinary forest, extending across the belly of South America, produces and recycles much of its own rainwater. In other words, most of the rain that falls on the expansive Amazon forest is actually generated by the forest itself. But as the Amazon rainforest is reduced in size through ongoing deforestation so, too, is its rainfall diminishing. This negative feedback loop is creating more droughts with associated wildfires throughout the Amazon.[35]

Consider the Oceans

It's typical to focus on the heat of the air around us when registering climate change, but it turns out that roughly 90 percent of the excess heat associated with global warming is accumulating in the ocean, with the ocean depths now warming almost ten times faster than was the case in the 1960s, 1970s, and 1980s.[36]

Of all marine habitats, coral reefs have been the most severely impacted by climate change. These reefs are built by, and made up of, thousands of tiny animals, related to jellyfish, known as "coral polyps." A reef begins to take form when a polyp attaches to the ocean substrate and commences to divide into hundreds of clones. As the polyps grow, they construct the skeleton-like body of the reef through the gradual deposition of calcium carbonate. Various types of algae sequester themselves in the bodies of polyps, receiving both the shelter and the carbon dioxide they need for photosynthesis. The polyps, in turn, benefit by receiving sustenance from their algal residents.

Though coral reefs have existed for millions of years, an estimated 20 percent have been lost since the 1980s and, given current trends, fully 60 percent could disappear by 2050.[37] Ocean acidification associated with climate change appears to be a big part of the reason. Indeed, fully one-third of human-generated greenhouse gas emissions have been absorbed by the sea. Once in the ocean, this added carbon dioxide is converted to carbonic acid, increasing seawater acidity.

Since the start of the Industrial Revolution—when humans began burning fossil fuels in earnest—the world's oceans have absorbed an estimated five hundred billion tons of carbon dioxide, thereby raising average ocean acidity by 30 percent. As seawater becomes more and more acidic, coral reefs cease their growth and gradually deteriorate.[38] The impact of this acidification on ocean fishes could also be catastrophic. Based on laboratory studies, oceanographer Eelco Rohling points out that given current greenhouse gas emissions trends, the acidity of the ocean will drop to a level "well beyond what fish and other marine organisms can tolerate with . . . serious implications for [fish] health, reproduction and mobility."[39]

Upshot: "Alter the chemistry of the ocean (as global warming is now doing) and the entire system shifts. Some natural changes we can predict, but it is impossible to anticipate how fast, or how much will occur as a consequence of tipping the ocean's chemistry onto a different course."[40]

Putting the Pieces Together

I have summarized the effects of climate change in selected regions of the globe—the Arctic, Antarctica, tundra, forests, oceans—but Earth operates as one interconnected whole. What happens in one place can, and often does, reverberate throughout the entire biosphere. For example, the disappearance of summer ice in the Arctic means that sea water that was once covered by ice is now being warmed by summer solar radiation, and this added warmth is beginning to contribute to the thawing of the Arctic permafrost with its huge stores of carbon-rich peat. As microbes break down this organic matter, releasing yet more greenhouse gases to the atmosphere, this contributes to warmer winters in places like the American West, thereby provoking more pest outbreaks, forest dieback, and forest fires and funneling yet more carbon to the atmosphere—resulting in yet more warming, triggering more melting of glaciers in places like Antarctica, potentially provoking sharp rises in sea level, and on and on.

In sum, climate scientists concur that Earth's climate is becoming increasingly chaotic. Now, here's the rub: Our entire civilization is built on the expectation of climate constancy. Sure, we can tolerate an occasional drought or hurricane, but our lives will begin to unravel if Earth's climate continues to shift in the direction of more frequent and unprecedented floods, hurricanes, droughts, wildfires, disease outbreaks, and sea-level rise. As writer and activist Bill McKibben reminds us: "Global warming is no longer a philosophical threat, no longer a future threat, no longer a threat at all. It's our reality."[41] Yes, whether we like it or not, we have become guinea pigs in a giant planetary experiment.

The Most Consequential Lie in Human History

The most disheartening thing about the unfolding of climate chaos is that none of this had to happen. It could have been avoided! The energy companies—so-called Big Oil—knew decades ago that fossil fuel burning could have disastrous consequences. Indeed, in 1977 one of Exxon's senior scientists, James F. Black, briefed many of the company's leaders, laying out the emerging research on what was then called the *greenhouse effect.* Black concluded his presentation by saying: "There is general scientific agreement that the most likely manner in which mankind is influencing the global climate is through carbon dioxide release from the burning of fossil fuels."[42] A year later, addressing another gathering of Exxon's executives, Black reported: "Independent researchers estimate that a doubling of the carbon dioxide concentration in the atmosphere would increase average global temperatures by 3.6 to 5.4 degrees Fahrenheit (and conceivably as much as 18 degrees F)."[43]

And just four years after this, in 1982, Exxon scientists, in a corporate document marked "not to be distributed externally," concluded that confronting global warming would "require major reductions in fossil fuel combustion." Otherwise "there are some potentially catastrophic events that must be considered," and to delay would be dangerous.[44]

Other fossil fuel conglomerates also knew what was happening. For example, in the late 1980s, Shell scientists were forecasting that atmospheric carbon dioxide levels could double as soon as 2030 and that "the changes may be the greatest in recorded history."[45]

But rather than taking action, Big Oil remained silent and continued on its business-as-usual path. In recounting this history of duplicity and immorality, Bill McKibben writes: "There should be a word for when you commit treason against an entire planet."[46]

What's *heartbreaking* is that it didn't need to come to this. Back in the 1980s and 1990s, those oil company leaders and their stockholders could have chosen a different path by summoning the dignity to act with integrity, declaring, in solidarity with climate scientists worldwide, that climate change was real and that urgent action was needed. In the long run, taking this stance might have actually benefited companies like Exxon and Shell, giving them a jump on creating a truly sustainable energy economy—that is, one heavily tilted toward solar and wind. Indeed, as far back as 1978, one of Exxon's managers said, "this [the specter of global warming] may be the kind of opportunity that we are looking for to have Exxon's technology, management and leadership resources put into the context of a project aimed at benefitting mankind."[47]

But instead of "benefiting mankind," the big oil companies resolved to benefit themselves by forming the Global Climate Coalition (GCC) in 1989 for the explicit purpose of undermining the emerging scientific consensus that climate change was a real and growing threat to human civilization. In Bill McKibben's words, this brazen action constitutes "the most consequential lie in history."[48] As a measure of Big Oil's success in duping the public, pollsters reported, as late as 2017, that roughly 90 percent of Americans still didn't know that there was an overwhelming consensus among climate scientists worldwide that climate warming, and the associated chaos, was no longer a theory, but a reality.[49]

Upshot: Big Oil's systematic disinformation campaign and its powerful lobbying efforts have played a key role in stalling climate action over the past three decades. In discussing Big Oil's callous behavior, environmental reporter Alex Steffen employed the term *predatory delay*—that is, "the blocking or slowing of needed change, in order to make money off unsustainable, unjust systems." Steffen went on to report: "If we had begun cutting global emissions in 1990, we could still have tackled the climate crisis with confidence."[50]

But we failed to act, and as a result, our future is now more uncertain than at any time in human history. It is true that our ancestors, back through time, also faced crushing hardships in the form of famine and plagues and war, but they lived secure in the assumption that humankind would persist on Earth—that is, that there would be future generations. Today, given what we have done, and continue to do, to Earth's climate, we no longer have this assurance.[51]

Foundation 5.2: Jeopardizing Earth's Natural Capital

When professionals in the business world use the word *capital*, they are referring to the instruments of *capitalism*: money, cash flow, investments, stock portfolios, profits, and so forth. Through this lens, Planet Earth is seen as a resource—that is, a product/commodity to be owned, bought, and sold. But consider that money is actually a placeholder for another form of capital that is more fundamental than money: *natural capital.* Natural capital refers to things like fertile soils, healthy oceans, robust forests, fresh air, and clean water—that is, the core elements that serve as the foundation for life on Earth. When Earth's natural capital is depleted or compromised, as is now the case, the well-being of life on Earth is jeopardized.

Jeopardizing Earth's Soil and Water Capital

Soil is alive; it is a source for the minerals that constitute our bodies; it is the basis for agriculture; it is precious. Under natural conditions, such as in a forest or a prairie, the land becomes more fertile over time, as countless generations of plants and animals flourish and then die, contributing their remains to the soil. Just as soil capital grows through these natural processes, it can also be depleted, as when topsoil is lost due to careless farming practices.

Ideally, over time new additions of soil capital will either balance or exceed soil losses. However, when more soil is lost each year than is formed, soil capital declines. Think of it as money in your bank account. If you are always taking money out and seldom putting fresh money in, you will eventually go broke. It is the same with soil!

The United States Department of Agriculture has been monitoring the status of U.S. soil capital for many years. Its data reveal that, on average, four tons of soil are lost per acre each year from U.S. croplands. In fact, U.S. soils are now eroding almost ten times faster, on average, than they are forming through natural processes. This depletion translates to an inch of topsoil loss, on average, from U.S. cropland every thirty years.[52] It is important to underscore the word "average." On some farms, the loss is greater than this; on others, less; and on some farms, there is actually a net gain of soil because of conscientious soil-building and conservation practices. Of course, if soil capital stocks were huge in the first place, it really wouldn't matter if a bit of soil was lost each year, but U.S. soil stocks are not huge. For example, already the farm state of Iowa has lost more than half of its fertile topsoil, and losses of similar magnitude are common in other farm states.[53]

In many parts of the world, average soil erosion rates are even higher than in the United States. For example, in China and India, countries containing roughly one-third of the world's population, soil is eroding thirty to forty times faster than it is forming; and worldwide, more than twenty-five million acres of cropland are being retired from farming each year due to the cumulative effects of soil depletion.[54]

IT'S NOT JUST SOIL CAPITAL THAT'S BEING SQUANDERED

Just as soil capital is being lost from U.S. farms, the same is true for freshwater capital. In the United States, the depletion of this natural capital is most dramatically evident in the Ogallala aquifer, an immense underground water reserve underlying the Great Plains. Currently, more than thirteen million acres of farmland in America's Great Plains are irrigated with water from this precious aquifer. New water enters the Ogallala each year in the form of rainfall that slowly percolates down through the soil to the water table, but irrigation water is being mined from the Ogallala much more rapidly than new water is being deposited. In fact, since irrigation began in the Great Plains in the 1940s, water levels of the Ogallala aquifer have been drawn down by more than one hundred feet; and at the current rate of extraction (2.7 feet/year), this aquifer may be mostly mined out within the next several decades.[55]

Jeopardizing Human Health

Just as fertile soil and abundant freshwater represent important natural capital stocks, a healthy, robust human being is also an important font of natural capital insofar as s/he can create and produce wonders. But human capital in the United States, when gauged by the metric of physical and emotional well-being, is now being compromised by poor diet, stress, sedentary lifestyles, depression, heart disease, obesity, cancer, diabetes, and respiratory ailments, along with a deteriorating natural environment.

Man-Made Chemicals Depleting Human Capital?

More than one hundred thousand different man-made chemicals are now in circulation, with an average of three to four new synthetic chemicals (with largely unknown health and environmental effects) released into the global environment each day.[56] Insofar as the vast majority of these chemicals have been invented within the last hundred years, there are now billions of tons of synthetic chemicals in circulation. Within our very bodies, we each now carry upwards of seven hundred chemicals that were not part of human body chemistry prior to 1900.[57] Many of these chemicals are classified as possible or probable cancer-causing agents (carcinogens).

The incidence of many human cancers has been rising. In some cases, this increase is partially the result of improved technologies for the early detection of cancers. For instance, prostate cancer can be detected more effectively now than in the past. But in the case of childhood cancers, which have increased significantly since 1970, early detection technologies are not employed and therefore are not a reason for the increases. Early detection is also not a factor in the threefold increase in testicular cancer for young men between nineteen and forty years old, nor is early detection a factor in the documented rise in brain cancers among both children and the elderly.[58]

The load of foreign chemicals that each of us carries in our body tissues—and by extension, the probability that any one of us might suffer from an environment-related cancer—depends, in part, on where we have lived during our lifetimes. Cancer-mapping studies often reveal a correlation between the abundance of certain categories of chemicals in the environment (e.g., pesticides, industrial solvents, radioisotopes) and the incidence of certain types of cancers. For instance, breast, bladder, and colon cancers are most common in regions that are highly industrialized (e.g., Eastern seaboard, Great Lakes basin, Mississippi Delta); while non-Hodgkin's lymphoma tends to be most concentrated in agricultural regions (e.g., Great Plains).[59]

If you are curious about the chemical loads in the place where you live, visit http://scorecard.goodguide.com/ and type in your zip code; this website will tell you the chemicals that are in your county's air, water, and soil—and by inference, in your body. When I visited Scorecard, I learned that my home county (Centre County, Pennsylvania) is among the dirtiest 30 percent of U.S. counties with regard to carbon monoxide, nitrogen oxide, and sulfur dioxide emissions and, by implication, overall cancer risk from hazardous air pollutants.[60]

The rise in human cancers over the past century has been paralleled by a rise in cancers among animals, especially since 1960. These animal cancers—for example, clams with gonadal cancer, fish with tumors—are almost always associated with contaminated environments, such as polluted bays and rivers.

Animal cancer studies are highly relevant to humans because many of the genes that govern cell division (i.e., those associated with cancer) are the

A DILEMMA FOR NURSING MOTHERS

The proliferation of foreign chemicals in the environment is taking some of the joy out of motherhood. Take the case of a woman who has just given birth to her first child. She knows about the advantages of breast-feeding—how it will fortify her newborn's immune system, helping to ensure that her infant will have fewer bouts of sickness. But what she hasn't heard much about is the chemical contamination of her breast milk.

When a mother nurses her infant, fat globules from throughout her body are mobilized and carried to her breasts, where they are transformed into milk. The problem is that toxins that have slowly accumulated in the mother's body fat over the years will also be part of those fat globules. Indeed, of all human food, breast milk is now among the most contaminated.

This presents today's mothers with what author and activist Sandra Steingraber calls an "obscene dilemma." What is a mother to do? Forgo the many advantages of breast milk by giving her child inferior infant formula? Or fortify her child's immune system with her own breast milk while simultaneously risking contaminating her baby with the synthetic chemicals stored in her own body fat? This is not a decision any woman should be asked to make.[61]

same (or very similar) in humans and other animals. Moreover, animal studies are easier to interpret than those involving humans because wild animals don't smoke, eat fatty foods, experience work-related stress, and so forth. In other words, animals don't introduce the confounding variables that make it so difficult to unequivocally pin down causality in human-cancer studies.[62]

Man-Made Chemicals Derailing Human Development

It was in the 1980s that a significant number of researchers began to entertain the possibility that synthetic chemicals—in addition to being possible carcinogens—could also interfere with human embryogenesis—that is, the process by which a fertilized egg in a mother's womb develops into a fully formed baby. Now, almost a half century later, there is unequivocal evidence that a surprising number of synthetic chemicals can inadvertently scramble the hormone signals necessary for the correct development of the human fetus.

Normally, as the fetus develops, hormones moving through the mother's circulatory system coordinate and direct the proper construction of the baby's body systems. Correct development depends on getting the right amount of the right hormone to the right place at the right time. If the hormone messages don't arrive, or if they arrive in the wrong amounts or at the wrong time, the embryo's development can be derailed (see figure 5.2).[63]

DEVELOPMENTAL ABNORMALITIES IN WILDLIFE

Numerous studies of amphibians, birds, and mammals have documented many types of developmental abnormalities, such as extra limbs in salamanders, shrunken penises in alligators, and eyes developing in mouth cavities in frogs.[64] Add to these the discovery that male fish—for example, small-mouth bass—in some U.S. rivers are now producing eggs. That's right: male fish producing eggs! Evidence is mounting that synthetic forms of the hormone estrogen may be a contributor to this so-called gender bending.

Fish biologists report that 80 percent of the streams tested in thirty states contain synthetic estrogen and its mimics. One source of these synthetic chemicals is chemical runoff from farms insofar as estrogen-mimicking chemicals are often present in the pesticides that are sprayed on U.S. farms. Only a tiny fraction of these sprays actually lands on the targeted pests; the rest ends up in farm soils that drain to waterways. A second source is the estrogen-like compounds that are common additives to soaps and cosmetics. A third source is the synthetic estrogens used in birth control pills and estrogen supplements. All of this eventually gets washed down drains and toilets on the way to sewage treatment plants and thence to surrounding waterways. Given this routing, it's not surprising that feminized male fish are especially common in stream stretches immediately below wastewater disposal facilities.[65]

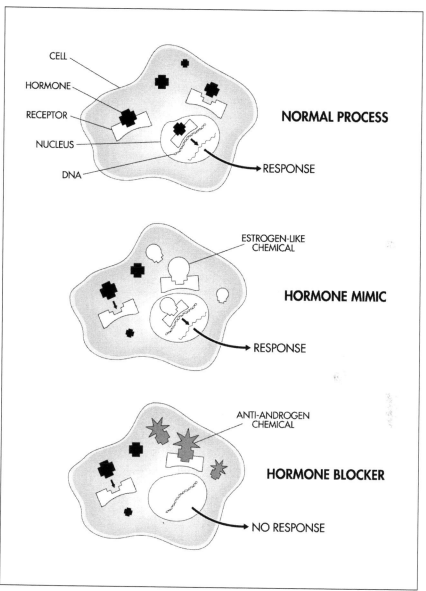

Figure 5.2. Hormones trigger cellular activity by hooking up—lock-in-key fashion—to hormone receptor sites, thereby activating genes in the nucleus (top cell). Foreign chemicals, like those that mimic estrogen, should they be present in the developing embryo, can bind with hormone receptor sites (e.g., middle cell) or block these sites (anti-androgen chemical—bottom cell), thereby disrupting the normal unfolding of embryo development.

Grasping the Ramifications of Endocrine Disruption

Scientists have already identified more than fourteen hundred chemicals—all man-made—that disrupt natural hormone function in humans.[66] These substances act by mimicking or blocking the action of the body's natural hormones. Such chemicals have been dubbed "endocrine disruptors" because they interfere with the normal functioning of the human endocrine system.

One hundred years ago, endocrine disruptors were virtually nonexistent. Now, these chemicals are showing up everywhere—for example, in detergents, toys, food-can liners, plastics, toothpastes, shampoos, flame retardants, pesticides, cosmetics, dental sealants, garden hoses, laptops—and by extension, in our very bodies.

A team of scientists reviewed eight hundred scientific studies on the effects of endocrine disruptors and concluded that it is "remarkably common" for very small amounts of hormone-disrupting chemicals to have profoundly adverse effects on human health.[67] This conclusion is not so surprising insofar as the endocrine system is the master control center of the body. Indeed, endocrine-disrupting chemicals can negatively impact the human reproductive system, thyroid signaling mechanisms, energy metabolism, fat cell development, and balance.[68]

The full significance of this endocrine disruptor saga is not completely understood as of yet. Some scientists believe that we are still seeing only the tip of the proverbial iceberg. For example, there may be connections between endocrine-disrupting chemicals and other health issues, such as the sharp increase in human obesity and diabetes in recent decades.[69] It also might be the case that endocrine disruptors are playing a part in the early onset of puberty, especially in girls. For example, breast development is now often starting at seven to ten years of age compared to ages thirteen to sixteen—the norm prior to the widespread introduction of endocrine-disrupting chemicals into the environment. Of course there could be other factors at play, but clearly something is amiss.

Decreased Fertility in Humans?

Of all the consequences of endocrine-disrupting chemicals, the significant decline in human fecundity over the past fifty years is perhaps the most worrisome. Multiple studies now reveal a correlation between endocrine-disrupting chemicals and disorders that result in decreased fertility in humans. For example, a recent meta-analysis (giant statistical analysis of dozens of research studies from scientists across the world) reported a 50–60 percent decline in sperm count in men from North America, Europe, New Zealand, and Australia between 1973 and 2011.[70]

These findings have led some scientists to hypothesize that events during male embryo development are at the root of the decline in the male sperm count. Here is one scenario that could explain what's happening. In the early months of male development, the emerging testes start to produce the hormone testosterone. Testosterone then triggers the development of the male

---

KNOWLEDGE IS POWER

If, after reading about all of this, you are motivated to learn where and how endocrine disruptors might be showing up in your life, you could start by doing an inventory of your personal care products. It's easy to do. Just enter the name of your shampoo (or deodorant or eye shadow, etc.) into the search window on the Skin Deep database (www.ewg.org/skindeep/). This will bring up Skin Deep's rating of your product, based on the known health effects of its ingredients. If your Skin Deep shampoo rating falls between 0 and 2, no need to worry; you are in the green. But maybe there are ingredients in your shampoo that fall in the 4–6 zone, or perhaps even the 8–10, red zone? In this case, it would be prudent on your part to carefully consider the health concerns associated with your shampoo. It could be that one of the ingredients is a carcinogen or a suspected endocrine disruptor or is known to produce allergic reactions. Find out. Knowledge is power.

---

testes while also fostering the growth of Sertoli cells, which later become the source for male sperm production. Given that adult males with lots of Sertoli cells are able to produce lots of sperm, while those with fewer Sertoli cells produce less sperm, biologists have speculated that for males with a low sperm count, something may have happened early, while still in their mother's womb, that limited Sertoli cell formation. For example, perhaps the pathway whereby testosterone activates Sertoli cell production was blocked by a foreign hormone-like chemical—that is, an endocrine disruptor—and this resulted in an adult male with fewer Sertoli cells in his testes and a correspondingly lower potential for sperm production.

This scenario is supported by research focusing on a class of hormone-disrupting chemicals known as phthalates. In lab experiments, scientists have shown that when female mice are fed small quantities of phthalates during their pregnancy, their male offspring produce abnormally low numbers of Sertoli cells, and upon reaching puberty, these same males exhibit reduced sperm production. Could phthalates be affecting human males in the same way? This is not implausible insofar as phthalates are common additives in many personal care products used by women, including cosmetics, fragrances, and nail polishes.

Stepping Back to See the Big Picture

In order for us to maintain our way of living, we must, in a broad sense, tell lies to each other, and especially to ourselves. . . . The lies act as barriers to truth. These barriers to truth are necessary because without them many deplorable acts would become impossibilities.

—Derrick Jensen[71]

We have reached a point where our power is so immense and our behavior so rapacious that we are destroying Earth, our home. We do this because—let's face it—we don't really care that much about Earth. It is our absence of a felt relationship with Earth that leads us to abuse her. If we cared, we would respect and honor her, like a beloved grandmother.

Wrap-Up: Expanding Ecological Consciousness

If you weigh the earth's terrestrial vertebrates, humans account for 30% of their total mass, and our farm animals for another 67%, meaning wild animals. . . . total just 3%. In fact, there are [now] half as many wild animals on the planet as there were in 1970, an awesome and mostly unnoticed silencing.[72]

We, humans, have become pretty good at numbing—going silent—when it comes to our mistreatment of Earth. For example, when a patch of Earth is clear-cut or strip mined or covered with concrete, or when the ocean is polluted or a coral reef destroyed, we sanitize what is happening with words like *resource extraction* or *growth* or *progress* rather than declaring the truth—that is, that we are killing, crushing, and destroying other beings, desecrating parts of the integrated whole that is Earth.

It seldom occurs to us that we are part of Earth's "family" and that, day by day, it is Mother Earth who extends hospitality to us, providing us with water to drink, air to breathe, and food for sustenance.

It's taken 4.5 billion years to create Earth's extraordinarily hospitable environment. Our ancestors—the microbes, the fishes, the reptiles, the mammals—have given us our body plan, our physiology, our genetic makeup. These ancestors have bequeathed us the biological equipment that now allows us to flourish.

But how dare we take any of this for granted! We are not rulers of Earth. On the contrary, we are guests! Just as it would be incredibly rude for a houseguest to plunder her host's home, our behavior has become an insult to Earth. We behave as we do because we fail to recognize that Earth is alive—that we are a part of her, and that what we do to her, we ultimately do to ourselves.

Forests and grasslands and wetlands and coral reefs and savannas are Earth's organs; soil is Earth's skin; the rivers and oceans are Earth's lifeblood. Extending this metaphor: All the species of animals and plants and microbes embedded in Earth's body are *her* cells, each in his/her own way connected to the greater whole that is Earth. It is the totality of species—the vast range of life strategies—that constitutes the magnificence of our Earth Mother.

We free ourselves from the hubris of speciesism, and the separation that it engenders, as we acknowledge that all species have their own unique sensitivities and that each sensitivity is a form of intelligence. For example, bees are attuned to things that you and I have no sensitivity to at all; the same is true for trees and squirrels and eagles—each with unique intelligences.

Even among humans, intelligence (i.e., sensitivity) manifests in individualized ways. For example, perhaps you are particularly sensitive to rhythms or to colors or to number patterns or to bird songs. Or maybe your intelligence is evidenced in your healing powers of touch or in your sensitivity to emotional nuance or in an ability to read lips. There are thousands of ways that human intelligence manifests, and most are only partially linked to cognitive function. The same is true for the millions of other species—each with their particular suite of intelligences—that have been spun into existence from Earth's ever-evolving body.

Homo sapiens Coming of Age

In many respects, we are still a species trapped in our adolescence, as revealed by our tendency to wage war rather than to make peace; to hoard rather than to share; to skirt the truth rather than to face it. Yes, we still have a lot of growing up to do, but might it be that today's tumultuous times are a perfect catalyst, patiently prodding us toward genuine adulthood?

Rather than framing our lives in terms of competition, control, and consumption—all of which tend to separate us from life—what if we choose to root our consciousness in love? What if we endeavored to love ALL BEINGS for themselves, not for their utility?[73] What if love is the revolution in human consciousness that will catapult us from our self-serving adolescence to a life-serving adulthood?

As our ecological consciousness expands, so, too, will our capacity for love, and hopefully a day will come when we behold all of Earth's beings as *all our relations*. On that day, the mere idea of mindlessly trashing Earth to make a quick buck will be thought of as obscene.

Applications and Practices: Courage

> In our modern culture men and women are able to interact with one another in many ways: They can sing, dance or play together with little difficulty but their ability to talk together about subjects that matter deeply to them seems invariably to lead to dispute, division and often to violence.
>
> —David Bohm, Donald Factor, and Peter Garrett[74]

We live in a mythic time. Never before, it seems, have humans wielded so much power for both good and harm. We can poison Earth and annihilate each other in war, but we also have the potential to embrace one another in peace and to restore Earth to wholeness. Do you believe this? If you don't think this is possible, you are not unusual. Chances are good that the person next to you doesn't believe it either. Indeed, many of us, around the world, have been conditioned to be spectators, allowing the media in their myriad forms to interpret the world for us.

We are told that we live in a democracy, as if simply being able to vote for candidates in elections orchestrated by *big money* constitutes a democracy. Our collective disenchantment and disempowerment are evidenced by the fact that, in the United States at least, many citizens don't even bother to vote. But believe it or not, everyday citizens have enormous latent power. The first step to uncapping this power is to muster the verve and courage to speak what we know to be true, first to ourselves and then to each other.

The Courage to Truthspeak

In the previous chapter we stressed that it is not enough to simply learn the facts about the beleaguered condition of Earth; it is also important to listen and, in so doing, to feel these facts—to take them in, somatically. This emotional digestion is a necessary prerequisite for accessing, and then speaking, our authentic truth in the face of Earth's decline.

Teacher and writer Tamarack Song calls on all of us to practice the art of "truthspeaking," which in his words means "to state clearly and simply what one thinks and feels. There is no judgment or expectation, no disguise of humor or force of anger. The manner of speech is sacred because it wells up from the soul of our being rather than from our self-absorbed ego."[75] Just the fact that you might pass a day without saying anything to another person that would be judged as a lie does not mean that you have practiced truthspeaking, because truthspeaking, at its core, is about not betraying— not lying to—ourselves.

Overall, truthspeaking is essential for the creation of a world grounded in courage, integrity, and wholeness. Like everything else in this book, it is in the doing that we learn and grow. So here are three practices for the cultivation of truthspeaking.

1. Truthspeaking by Treating Each Word as Special

You have no doubt heard the expression that *less is more*. This phrase seems counterintuitive. Really, how could less be more? Perhaps, for starters, call to mind people you have encountered in your life who speak few words but offer great wisdom. Compare them to those who ramble and rattle on in endless monologues.

You can explore the truth in *less is more* for yourself by engaging in an exercise called "If I Had Only Five Words." The rules are straightforward.

Sit facing another person and simply take turns directing questions to each other and responding. Avoid questions that are superficial; instead, ask questions that invite deep and soulful responses. Now, here's the critical part: You only have five words to answer each other's questions. It is not necessary to use complete sentences. For example, if your partner asks, "What makes your heart ache about being alive in these times?" wait with curiosity and patience until five words emerge from the core of your being that are your truth. If you muster the courage to genuinely engage in this practice, you will discover, if you haven't already, that less really can be more.

2. Truthspeaking to Ourselves

In my experience, it is exceedingly rare to find a human being who doesn't pretend in some ways. Often the pretending I do is subtle, meaning that I may not even be aware that I am pretending. How about you? For example, do you sometimes pretend:

- You have it all together when you really don't.
- You like certain people when you really don't.
- You know things when you really don't.

You can explore this human tendency to pretend by completing, three times, this open sentence:

I pretend that _____.

I pretend that _____.

I pretend that _____.

Next, pick the statement that is the most troubling to you. For example, perhaps you pretend to like someone whom you don't really care for. Circle the statement—whatever it is. Then, consider the following four questions:

1. How does pretending in this way serve you? After all, you wouldn't be pretending unless you *believed* it served you (benefited you) in some way!
2. What is the personal cost you pay for pretending in this way?
3. In what ways might your life change for the better if you stopped pretending?
4. What is a new possibility—a new way of being—that would allow you to move beyond pretending and toward authenticity?

None of this is easy, but being honest with ourselves is a gateway to freedom—especially freedom from self-deception.

3. Truthspeaking for a Day

When you are ready to dip your toes deeper into the waters of truthspeaking, have a go at truthspeaking for an entire day. Begin your day with fifteen minutes of silence. Then, when the inclination to speak arises, simply ask yourself, "What is true for me right now?" Ask it and mean it. Ask it and then go inside yourself for the answer: What is true in your head? What is true in your body? What is true in your heart?

Though many of us are not very practiced at speaking from a place of embodied presence, we can cultivate this skill. For example, pause right now and ask yourself, what is true for you in this very moment? Go slow: What's true . . . right now . . . for you? Be patient. Be curious.[76]

As an incentive for this practice, imagine a world where each of us has the resolve to speak what is true in all realms of our lives, including voicing our truth regarding *the unraveling of the biosphere*. It is this courage to stand and fearlessly give voice to our truths that will determine not just our own futures, but the future of humanity.

Questions for Reflection

- What eyewitness evidence can you offer from your own life regarding climate change?
- Imagine that because of catastrophic climate change, starting tomorrow you have to cut your energy usage by half—that is, you will have to use only half the electricity, half the gas, half the heat, half the living space, half the stuff that you are accustomed to using. In what ways would your life be diminished, as well as enhanced, if this came to pass?
- Based on your exploration of Scorecard (http://scorecard.goodguide .com), what is the situation regarding chemical pollutants in the place where you live, and how does this information sit with you?
- Humankind can survive without access to capital in the form of money, but without access to *natural capital* we are all doomed. Do you agree or disagree with this statement? Why?
- What does it feel like for you to be a kind of *guinea pig* in a largely unplanned planetary experiment involving possible chemical havoc and catastrophic climate change?
- In what ways do you avoid truthspeaking? What's holding you back? What benefits might accrue to you, and to those around you, if you were to fearlessly speak your truth?

6

Living the Questions

Discovering the Causes of Earth Breakdown

> Once you have learned how to ask questions—relevant and appro-
> priate and substantial questions—you have learned how to learn,
> and no one can keep you from learning whatever you want or
> need to know.
>
> —Neil Postman and Charles Weingartner[1]

I have directed most of my work in environmental science toward assessing the health of Planet Earth. In the process, I've asked lots of questions. Sometimes these questions have involved field research. For
example, in the mid-1970s, noting that large swaths of Amazon basin rain
forest were being hacked down and turned into cattle pastures, I wondered
why this was happening and if it would ever be possible for forests—complete with native trees and wild animals—to reestablish on denuded Amazon pasture lands. I spent the better part of two decades pursuing answers
to these twin questions.[2]

This chapter is centered on the all-encompassing question: What are
the root causes of the ecological upsets described in the preceding chapters,
including deforestation, species extinction, ocean degradation, and climate
disruption, as well as the widespread chemical pollution of Earth's air, water,
and soil (and by extension the contamination of our very flesh and blood).
Throughout my research, I have not been satisfied to simply learn what has
been happening—I want to know why! This quest to understand *the why of
things* is akin to peeling back the layers of an onion, promising the possibility
of deeper understanding as each successive layer is stripped away.

The transformative power of "why" is illustrated in a story about
Primo Levi, a Jewish man who was forced to endure a long journey in a
cattle car destined for a concentration camp during World War II. Deep
into the journey, the train lurched to a stop and Levi, who was severely
dehydrated, spotted an icicle hanging from a bridge abutment. He reached
out to break it off, but before he could bring the icicle to his lips, a hulking

guard grabbed his arm and snatched it away. Dismayed, Levi looked at the guard and asked, "*Warum?*" ("Why?"). The guard responded: "*Hier ist kein warum*" ("There is no '*why*' here").[3]

Banishing "Why?" from our lives is tantamount to extinguishing the core of our humanity. Why? Because the search for meaning is fundamental to human existence and there is arguably no better lever to guide us in this search than the word "Why?"

Foundation 6.1: Humans—Too Many, Too Much?

As we reach the quarter point of the twenty-first century, Earth's human population is growing by approximately 9,000 people per hour, 225,000 people per day, and 1.6 million people per week. At present (2020) the human population is tipping the scales at 8 billion, and if projections are correct, there will be 10 billion of us by 2050.[4]

But get this: It took until 1850 for the human population to reach one billion people. Though we had been around as a species for some two hundred thousand years, it had taken all of that time to get to one billion! But then, in a matter of only eighty years, our population doubled to two billion. That's right; in the eighty-year period from 1850 to 1930 we did what it had taken us two hundred thousand years to achieve. And we didn't stop at two billion: In the forty-year span from 1930 to 1970, our population doubled again, from two to four billion; and now we are doubling yet again, to eight billion.

If you are following me, you may be wondering how it is even possible for a population to double in size so quickly. To understand, imagine an island with a population growth rate of 5 percent per year. If this hypothetical island has 2,000 people at the beginning of the year, its population will have grown to 2,100 people by the end of the year ($0.05 \times 2,000$ people = 100 new people added to the population). After two years the population will have grown to 2,205 people ($0.05 \times 2,100$ people = 105 new people). If you continue with these simple calculations, you will discover that in fourteen years the island's population will have grown to four thousand people. In other words, given a 5 percent population growth rate, the population will have doubled, from two thousand to four thousand, in just fourteen years! And, given fourteen *more* years, this hypothetical population would double again, to 8,000 people; as long as the growth rate stayed at 5 percent, the population would continue to double every fourteen years.

Using the so-called Rule of 70, it is possible to quickly calculate the doubling time of any population by dividing the number 70 by the percent growth rate. Hence, in our example, $70/5\% = 14$, the number of years it took for the island population to double from one to two thousand.

So far we have addressed the *how* of population growth, but not the *why*: Why has our human population been exploding since the 1800s? If you are like most folks, you might assume the reason is simply that women are having more babies these days than in the past. But questioning assumptions is important—especially theories like this that may appear self-evident.

It turns out that, based on extensive demographic research, the recent surge in human population growth has been primarily the result of a decline in human death rates, *not* an increase in birthrates. In the centuries prior to the 1700s, women were still birthing roughly the same number of babies, but starting in the 1800s, proportionately more of those newborns were surviving the challenges of birth and early childhood. The same was true for mothers; that is, childbirth became less risky for mothers. The primary explanation for this change is modern medicine, especially the realization that sanitation measures (as simple as hand washing) could greatly reduce the spread of deadly microbe-borne infectious diseases. A second factor leading to increased survival was the development of vaccines, particularly those aimed at ridding the world of potentially lethal childhood diseases like measles, whooping cough, polio, and diphtheria. More babies surviving to adulthood meant more mothers and, in turn, more babies.

As this edition goes to press, our population growth rate is hovering at 1 percent per year. Given this rate, we can use the Rule of 70 to calculate how long it will take for our numbers to double from eight billion

THE WHY BEHIND HIGH POPULATION GROWTH

Among those living in "developed" countries (e.g., United States, countries of Western Europe, Japan), there is a tendency to blame those in "less-developed countries" (LDCs) for the human population explosion. While it is true that people in LDCs often have large families, have you ever wondered why? One way to explore this is to imagine yourself as a member of a cash-poor family living in the highlands of Ecuador. You are eighteen years old, the eldest of five children. Your family has no electricity and relies on wood for cooking and heating. Your days, and those of your siblings and parents, are spent engaged in such essential activities as creating and repairing terraces on the steep slopes where you plant your crops; cultivating, weeding, and harvesting vegetables and grains; processing and preparing food; herding goats; making cheese; weaving fabrics for your clothes; constructing and repairing tools; gathering wood and water; and caring for each other in times of sickness. As a member of this large hardworking family, you would know, instinctively, that you were able to survive because there were many hands and legs and backs and heads to help with the many time-consuming tasks necessary to run a self-sufficient household.

to sixteen billion. The answer is 70 years ($70/1\% = 70$). This estimate is based on the assumption that our population growth rate will remain at 1 percent throughout the rest of this century. However, our growth rate has been declining from a high of 2 percent/year in the 1960s (Think: thirty-five-year doubling time!) to its present rate of 1 percent, and most demographers predict that the growth rate will continue to decline, with human numbers topping out at around eleven billion toward the end of this century. We shall see. In the meantime, it is heartening to note that there are already some countries—for example, Japan, Italy, Hungary, Greece, and Russia—where population size has actually begun to decline (i.e., where population growth rates have turned negative).

It's even possible that at some point the human population, as a whole, will begin to decline. For example, if all adult couples on Earth suddenly resolved to limit their family size to just one child, Earth's human population would sink back to below one billion within two-hundred years—that is, within roughly the same span of time it has taken to go from one billion (1850) to eight billion (today).

The Instructive Case of Kerala

It has long been noted that families in LDCs tend to be larger than families in more developed countries (see box), but there are fascinating exceptions to this pattern. Take the state of Kerala in southern India. The citizens of Kerala are poor—that is, their per capita income mirrors that of LDCs—but that's where the similarities end. In spite of its low per capita income, Kerala looks almost identical to the United States in terms of key quality-of-life indicators such as life expectancy, birthrate, and literacy (see table 6.1).[5]

Table 6.1. Quality-of-Life Indicators for Less-Developed Countries, Kerala, and the United States

Quality-of-Life Indicators	Less-Developed Countries	Kerala State, India	United States
Per capita income ($/year)	400	350	32,000
Life expectancy (years)	58	76	72
Birthrate (#/1000/year)	40	16	14
Literacy (%)	50	94	96

Sources: United Nations Population Fund (UNFPA), "Population Dynamics in the LDCs: Challenges and Opportunities for Development and Poverty Reduction," www.unfpa.org; Wikipedia, s.v. "List of countries by population growth rate," http://en.wikipedia.org/wiki/List_of_countries_by_population_growth_rate; Wikipedia, s.v. "Kerala," http://en.wikipedia.org/wiki/Kerala#Human_Development_Index; "Population Development," Soapboxie, January 31, 2019, https://soapboxie.com/world-politics/Population-Development-What-Kerala-can-Teach-India-and-China; and Kimberley Amadeo, "Income Per Capita, with Calculations, Statistics, and Trends," The Balance, October 4, 2019, https://www.thebalance.com/income-per-capita-calculation-and-u-s-statistics-3305852.

Why does Kerala seem to defy the rules of poverty? Science writer Bill McKibben, who traveled to Kerala to unpack this paradox, frames the mystery this way: "Kerala undercuts maxims about the world we consider almost intuitive: Rich people are healthier, rich people live longer, rich people have more opportunity for education, rich people have fewer children. We *know* all these things to be true—and yet here [in Kerala] is a counter case. . . . It's as if someone demonstrated in a lab that flame didn't necessarily need oxygen, or that water could freeze at 60 degrees. It demands a *new chemistry* to explain it, a whole new science."[6]

Among the factors that help to explain the paradox of Kerala, the most significant appears to be the respect that is accorded to women in Kerala. Typically, women in LDCs are subjected to systemic discrimination insofar as they receive less education, own less land, have less political power, get paid less money, have less free time, and exercise less control over their reproductive lives compared to their male counterparts. Indeed, Kerala's "new chemistry" has ingredients that are rarely seen in LDCs: equal rights for women, redistribution of wealth, and equal access to education and health care.[7]

The education now afforded to all of Kerala's citizens has been especially important for women, empowering them to take charge of their lives. As a reflection of this, the average woman in Kerala doesn't marry until twenty-two years of age, versus eighteen in the rest of India. As a further testimony to the respect accorded to women, Kerala is the only place in all of Asia where young girls outnumber young boys—a strong indication that gender is not a factor in abortions.[8]

Reframing Problems as Opportunities

We frame a painting to give it definition so that it stands out from its surroundings. If we choose the wrong size frame—too small or too large—the picture doesn't look right. In a similar way, questions are often used to frame problems, but if our questions are cockeyed—for example, if they are laden with assumptions—they will interfere with our ability to answer the question before us. Ellen Langer, in her book *Mindfulness*, makes this point with the following story:

> Imagine that it's two o'clock in the morning. Your doorbell rings; you get up, startled, and make your way downstairs. You open the door and see a man standing before you. He wears two diamond rings and a fur coat and there's a Rolls Royce behind him. He's sorry to wake you at this ridiculous hour he tells you, but he's in the middle of a scavenger hunt. . . . He needs a piece of wood about three feet by seven feet. Can you help him? In order to make it worthwhile, he'll give you $10,000. You believe him. He's obviously rich. So, you say to yourself, "how in the world can I get this piece of wood for him?" You think of the lumberyard; you don't

know who owns the lumberyard; in fact, you're not even sure where the lumberyard is. It would be closed at two o'clock in the morning anyway. You struggle but you can't come up with anything. Reluctantly, you tell him, "Gee, I'm sorry."

The next day, when passing a construction site, you see a piece of wood that's roughly three feet by seven feet—it's a door! You could have just taken one of the doors in your house off its hinges and given it to that rich man in return for $10,000. Why on earth, you say to yourself, didn't it occur to me to do that? The reason, of course, is because yesterday your door was not a piece of wood. The seven-by-three-foot piece of wood that the man was searching for was hidden from you, stuck in a category [frame] called "door."[9]

As illustrated by Langer's story, when we fail to recognize the categories that limit our seeing, we become puppets to our conditioning. For example, the population problem was for many years framed as a problem of ignorance—that is, poor people were having too many children because of ignorance! Given this narrow frame, the solution was birth control—plain and simple. Fortunately, today population growth is coming to be understood in the context of a deeper set of issues rooted in systematic discrimination against women in social, educational, political, and economic realms. With this shift in understanding, the question is no longer "How can women be prevented from having more children?" (a narrow and sexist frame), but "How can women be empowered in ways that enhance their freedom, autonomy, and status?" As reforms are enacted in these new directions, women, of their own accord, choose to have fewer children, as documented in Kerala.

The Population-Consumption Link

It would be easy to conclude, given the recent explosion in human numbers, that population growth is at the root of all the ecological problems that humankind now faces, and that if we could simply stabilize our population, everything would be OK. What do you think? Would that do it?

Before answering, consider that the dramatic rise in human numbers in recent times has been mirrored by an even sharper rise in the production and consumption of consumer goods—for example, toasters, fertilizers, plastics, cars, paper, guns, cell phones, computers, shoes, and hundreds of thousands of other items (see figure 6.1).[10] This surge in the production of *consumer goods* has been accompanied by all sorts of *bads*, including enormous waste, freshwater and ocean contamination, and climate chaos in addition to significant declines in forest cover, soils, ocean fisheries, and freshwater aquifers. Economists tend to classify these bads as "externalities," but they aren't "external" at all because they are undermining the health and well-being of Earth and her inhabitants.

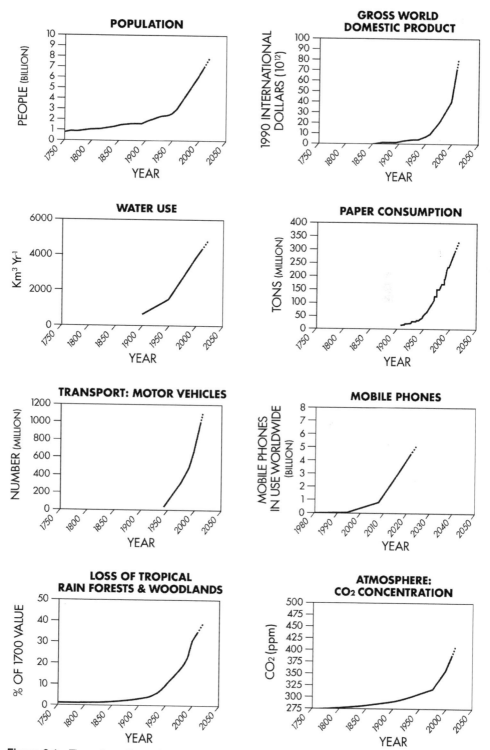

Figure 6.1 The extraordinary rise in human population, together with sharp increases in consumption and environmental impacts.

The Consumer Lifestyle

The rate at which humankind is now consuming Earth's resources greatly exceeds the rate of human population growth. Therefore, even if human population growth ceased starting today, our collective human impact on Earth would still grow because each of us, on average, is consuming more and more stuff each year. For example, in the case of the United States, per capita consumption is three times greater now than in 1960.[11]

All the stuff that we consume, of course, comes from somewhere, and that "somewhere" is *Earth*. In 1970 humankind, as a whole, consumed an estimated twenty-two billion metric tons of Earth materials, consisting mainly of metals, fossil fuels, biomass, and minerals. By 2010, four decades later, human consumption of Earth materials had increased more than three-fold, to seventy billion tons. Given that the global human population was seven billion in 2010, consumption of Earth resources averaged twenty thousand pounds per person.[12]

These big numbers are difficult to comprehend, so let's break it down and consider just the United States in the year 2000. In that year, each American, on average, consumed 132,000 pounds of resource-related materials, consisting of oil, sand, grain, iron ore, coal, and wood. Even this number is hard to grasp, so try this: In 2000, an average American was consuming an eye-popping 362 pounds of natural resources each day—15 pounds each hour! And that was back in 2000![13] Since then, material consumption has continued to rise.

If you'd like some eyewitness evidence of this growth in consumption, consider the rise of storage lockers in the United States. This industry was essentially nonexistent prior to 1970, but since then the business of storing stuff has grown so precipitously that storage lockers now occupy a cumulative area of roughly seventy square miles—equivalent to twenty-one square feet of storage space for every household in America.[14] But the bulk of what Americans purchase these days isn't stored at all, because most of the stuff we buy is used for a short time and then thrown away. For example, in just one year (2016), Americans discarded more than 152 million cell phones, each containing hazardous substances, such as lead, arsenic, and cadmium, that are associated with some of the same cancers and reproductive abnormalities discussed in chapter 5.[15] Though recycling options are now available, the majority of old cell phones end up in landfills, where their toxins slowly leach into Earth's soil and water. The same story—that is, easy come, easy go—applies to most of the other things that Americans purchase. Is it any wonder that U.S. media venues routinely refer to Americans not as *citizens* but as *consumers*?

Why We Consume

Humans are biologically predisposed to be attracted to snazzy stuff, but our current *hyper-consumer* lifestyle didn't just happen as a natural result of our biology. Our predisposition to consume is a cultural phenomenon as much as anything—the result of the particular economic system in which we live.[16]

Marketers in the United States spend in excess of $200 billion a year (equivalent to $600 per person/year) to convince us that *we* need what *they* have to sell.[17] No surprise, then, that many of us find ourselves buying products that we don't really need. For example, today more than half of us in the United States drink bottled water. In other words, we pay top dollar for something that would only cost us a fraction of a penny if we accessed it from a household tap. When we purchase water in plastic bottles at a store, what we are mostly paying for is the plastic bottle and the cost of transporting that bottle to the store. Indeed, the bottled water we buy often comes from hundreds, even thousands, of miles away. All this transport, combined with the manufacture of the plastic bottles, requires fossil fuels; for example, approximately a half cup of oil is required to produce a one-liter plastic water bottle, and ironically, roughly two liters of water are required to manufacture the bottle.[18] Here is a further irony: Roughly half of the bottled water sold in the United States, including popular brands like Aquafina, Dasani, Evian, and Fiji, is actually sourced from the tap.[19]

Some years back the Natural Resources Defense Council (NRDC) tested the quality of more than a hundred brands of bottled water and found that one-third contained significant impurities (e.g., levels of chemical or bacterial contaminants exceeding those allowed under state or industry standards). In fact, the NRDC study reported that household tap water is subject to more stringent regulations than is the case for bottled water.[20]

And what happens to all those plastic bottles after the water is consumed? The majority are discarded and end up in landfills, where they will require thousands of years to break down. As for the bottles that get recycled, there is a good chance, these days, that they will be exported to other countries for resource recovery.[21]

I have singled out bottled water for illustrative purposes, but there are tens of thousands of other products put on the market each year, each one whispering: "Buy me!" You need only stroll up and down the aisles of a nearby mall or a superstore to experience the pervasive calls of marketers

---

SOME COLLEGES SAYING "NO" TO BOTTLED WATER

Given the questionable wisdom of bottled water as a commodity, some student activists have prodded their colleges to institute policies banning the sale of bottled water. As one Dartmouth student observed, "The product just doesn't make common sense. Companies are taking something that is freely accessible to everyone . . . packaging it in a non-reusable container, and then selling it under the pretense that it is somehow better than tap water."[22]

enticing us to buy more stuff. But what if, before making *any* purchase, you were to pause and ask questions like these: Could I live without this product? Will this item truly enhance my well-being? What about the well-being of Mother Earth? When I throw it away, where will it go?

When I did this recently at my local big-box store, I was amazed to realize that most of the things lining the shelves didn't even exist as product categories when I was growing up. In other words, I was being tugged to purchase stuff that until recently had never existed and that both I and the world could easily live without.

Upshot: Our present environmental crisis is not simply the result of rapid human population growth. Consumption, it turns out, is an even more critical factor. Each item that we buy—whether it's a car, a pack of gum, or an iPod—comes from Earth and goes back to Earth, leaving a trail of impact in its creation, use, and disposal. What's to be done?

Foundation 6.2: First-Person Ecology: Our Ecological Footprint

We live in a remarkable historical epoch: population rising, consumption climbing, technology expanding . . . and Earth hurting. Can it continue? Does Planet Earth have enough wealth for all of us to consume as much as we want for as long as we want?

Fortunately, there is a metric called the "ecological footprint" that has been explicitly devised to address these questions. Your ecological footprint is the area of productive land and sea that is required to produce all the things that you consume each year plus the area necessary to absorb all the waste associated with your consumption.

Everything we consume has a footprint. Food is one part of our footprint; some other categories are clothing, housing, furniture, appliances, and electronics. Add to this our energy needs for heating, cooling, electricity, and transportation.[23] And don't overlook all of the little things that we purchase day after day. Everything on store shelves, without exception, has a footprint.

In practice, calculating ecological footprints is an accounting exercise. By way of example, consider a quarter-pound hamburger that you might eat for dinner. The beef patty came from a steer. During his life, that steer required land to graze on and later, when he was being fattened, land was needed to grow the corn and grain that the steer consumed. When fully grown, he was slaughtered at a meat-packing plant. Both space and energy were required for the plant's operations, and additional materials and energy were needed to process, pack, and transport the beef, via refrigerated truck, to a grocery store or restaurant for marketing. These sales venues are also a part of the burger's footprint, insofar as energy and materials were used in their construction and daily operations. By piecing together all the steps involved in producing and delivering the burger to your belly, it is possible to estimate the land and water required for its production along with the carbon/energy footprint, expressed as carbon dioxide released to the atmosphere. Here's the tally for one quarter-pound burger: water = 14 gallons; corn/grain feed = 13 pounds; land = 65 square feet; carbon release = 4 pounds.[24]

The Power of Footprinting

By means of *footprinting*, we are now able, for the first time, to determine the capacity of a region or nation or planet to support its human population. The prospects, from good to grim, depend on the amount of productive land/sea relative to the number of people and the amount that those people are consuming.

As summarized in table 6.2, the total productive acreage on Earth is 32.3 billion acres. This total is the size of the Earth *pie* available to humankind to meet our material needs.

But wait a minute! What about the millions of other species that cohabit Earth with us? They also need productive land and sea to flourish. Taking this into account, the World Commission on Environment and Development recommended that a minimum of 12 percent of Earth's productive land/sea surface (i.e., 3.9 billion acres) be set aside, as parks and wildlife preserves. This leaves humans with a 28.4-billion-acre pie to meet our needs (table 6.2). The slice of this pie that each individual person requires to support his/her lifestyle constitutes that person's ecological footprint.

What's your guess about the size of your ecological footprint? Less or more than an acre? You can get a personalized answer by going to http://myfootprint.org/en/your_carbon_footprint/ and filling out an ecological footprint quiz.

Insofar as consumption levels vary from country to country, average footprint sizes also vary by nation. For example, the average ecological footprint of a U.S. citizen is twenty acres. That's about twice the size of the footprint of the average European and eight times more than the average citizen of

Table 6.2. Productive Acreage on Earth (in Billions of Acres)

Cropland	3.7
Pasture land	8.5
Forest	12.8
Productive ocean surface	7.3
Total	32.3
Adjusted total	28.4[a]

[a] Assumes setting aside 12 percent of productive land/sea in parks and preserves.

India.[25] Within any given country, the wealthier a person, the higher his/her footprint tends to be, as consumption is correlated with income. For example, in the United States, someone earning $200,000 per year might have a footprint greater than fifty acres, while someone taking home only $25,000 per year would in all likelihood have a footprint in the ten-acre range.

Footprinting can also be used to assess a country's vulnerability to economic threats. For example, knowing a nation's population size and its average per capita footprint will reveal if that country has enough productive acreage within its boundaries to support its population. Some countries (e.g., Brazil, Australia) have huge productive capacity and relatively small populations. Given their ecological assets, these countries have better prospects for withstanding economic crises than high-population countries with limited productive capacity. Countries like Singapore, for example, with sparse ecological assets and high population densities, have to rely on productive acreage outside their borders to meet their population's needs. Overall, of the 153 countries where footprint assessments have been completed, 107 don't have enough productive acreage within their borders to supply their people's needs.[26] The United States falls in this category. We are a debtor nation in ecological terms, because we don't have enough land and sea within our borders to satisfy our present consumption demands.

The U.S. situation reflects that of Earth as a whole. Indeed, humanity's composite global footprint now exceeds Earth's productive capacity by about 50 percent, meaning that it now takes eighteen months for Earth to regenerate the resources that humankind consumes in twelve months. Upshot: Humans now need, at a minimum, 1.5 Earths to sustain their current levels of consumption.[27]

But wait a minute. How can this be? How are we able to survive if we require one and a half Earths to supply our needs but only have one Earth available to us? To deconstruct this paradox, consider, first, that Earth still has considerable surplus stocks of natural capital (see chapter 5) in the form of soil, water, forests, fisheries, and minerals. However, since the mid-1800s, humankind has been steadily depleting the water in Earth's aquifers, the soil in Earth's crop and agricultural lands, the wood in Earth's forests, and the fishes in Earth's seas. Simultaneously, we have been putting enormous amounts of waste into the biosphere and, in the process, contaminating Earth's soil, water, and atmosphere, all the while undermining the stability of Earth's climate. It is now scientifically clear that if we persist in this imbalance of taking and wasting, we will soon exhaust Earth's natural capital surplus, making ecological breakdown inevitable.

The ecological crisis that now envelops us was foreseen back in the early 1970s when scientists at the Massachusetts Institute of Technology provided compelling evidence that unbridled growth in human population, consumption, and waste generation would lead to the degradation of

Earth and ultimately to a sharp decline in human well-being. The analysis behind these predictions was laid out in the book *The Limits to Growth*.[28] Though the MIT researchers weren't right on every count, their overall analysis was spot on.

Today, the evidence that Earth's biological support systems are severely compromised is undeniable, yet it seems that many of us remain mired in denial. How about you? If you lean toward denial, consider this statement by renowned ecologist Jared Diamond, who researched the breakdown of past civilizations: "One of the main lessons to be learned from the collapses of the Maya, Anasazi, Easter Islanders and those other past societies, is that a society's deep decline may begin only a decade or two *after* the society reaches its peak numbers, wealth and power."[29] This makes ecological sense insofar as peaks in population, wealth, and consumption (such as those that we are now witnessing) correlate with increased impacts on Earth's biological support capacities.

Stepping Back to See the Big Picture

I used to think the top environmental problems were *biodiversity loss, ecosystem collapse* and *climate change*. I thought that with 30 years of good science we could address those problems. But I was wrong. The top environmental problems are *selfishness, greed* and *apathy* and to deal with those we need a cultural and spiritual transformation.[30]

—Dr. Gus Speth

Ecological footprinting provides a lever for considering some important ethical questions. For example, who, in your view, bears more responsibility for the compromised condition of Earth: a couple from India with six children or an American couple with two children? Certainly six children seems a bit excessive—even irresponsible—but look again, this time through the lens of *footprinting*. Given that the average citizen of India has a footprint size of only 2.5 acres, that large Indian family has an aggregate footprint of 20 acres ([2 adults + 6 children] × 2.5-acre footprint each = 20 acres). A 20-acre family footprint may sound like a lot, but the Indian family's footprint is actually four times less than the 80-acre footprint of the two-child American family [(2 adults + 2 children) × 20-acre footprint each = 80 acres). Upshot: From the perspective of Earth, it is not family size that matters so much as the piece of Earth's pie that each family member takes. Viewed through this lens, people living in developed countries, like the United States, have a more serious *population* problem—that is, they exert a far more significant impact on Earth—than people living in LDCs, like India. This is not about assigning blame so much as being honest with ourselves and each other.

Here's a second example of how footprinting helps to raise ethical issues. It starts with the question: How much productive acreage would be available for each human inhabitant of Earth if we all received an equal share? We can determine this amount by simply dividing human population size (roughly 8 billion) into the productive surface area of Earth (28.4 billion acres; table 6.2). The answer comes to 3.5 acres; this would be a *fair-earth share*, assuming Earth's resources were divided equally among all human beings.

Now consider that the current U.S. per capita footprint of 20 acres is more than five times a fair-earth share. Indeed, if everybody on Earth had a U.S.-sized footprint, Earth could only support 1.4 billion people (28.4 billion productive acres/20 acres per person = 1.4). Conversely, if all the humans on Earth had a U.S.-size ecological footprint, five Earths would be necessary support us.

All of this footprinting leads to some important insights, providing us with a tool for candidly viewing our situation through both an ecological and a moral lens. Indeed, footprint assessments are like bank statements, and bank statements are helpful tools—even when they tell us that our finances are in disarray—because they provide the information and, hopefully, the motivation to set things right.

Setting Things Right

Consider, if you dare, what it would mean for a U.S. citizen to live within the bounds of a 3.5-acre fair-earth share. Would it even be possible for us to reduce our footprint fivefold, from 20 acres to 3.5 acres? I recently met with some friends to explore this question. We began by looking at the things we consume and then considered footprint reduction strategies for those things. We started with our personal computers. The options we came up with included: (1) don't bother to regularly replace "old" computers with brand-new ones and instead support a local business to upgrade outdated computer components as needed; (2) team up with a friend and share a computer; and (3) forgo owning a computer and instead use computers available in public libraries.

We went on to discuss other categories of consumption—for example, clothing, housing, entertainment, transportation—and in so doing, came to see that for each category, viable approaches for reducing our footprint size as much as fivefold were, in fact, available. Later I realized that there was a bigger question undergirding our conversation that day: Is it possible to reduce our ecological footprint while at the same time increasing human well-being? Let's have a look.

Based on global surveys by sociologists, human beings all over the world say that what they want most in life—for themselves and those around them—is happiness. People also mention money, but the longing for happiness is more prevalent than the longing for wealth. With these results in mind, researchers at the New Economics Foundation (NEF) have devised an index of progress called "The Happy Planet Index" (HPI). Rather than using conventional measures, such as gross national product, to assess national well-being, NEF focuses on people's happiness, believing, quite sensibly, that a successful nation is one inhabited by happy people. One encouraging outcome of the NEF's research is the finding that it is possible for people to have small ecological footprints *and* still be happy. In other words, small footprints and happiness are not mutually exclusive.

Costa Rica ranks number one on NEF's Happy Planet Index insofar as its citizens, on average, have high levels of happiness with small footprints—that is, they make very modest demands on Earth's resources. By contrast, the United States, with a per capita footprint three times that of Costa Rica, ranks a lowly 108th on the Happy Planet Index.

For countries (and individuals) interested in improving their Happy Planet Index scores, research reveals that happiness can be enhanced through five simple actions: (1) connecting with others, (2) increasing physical activity, (3) spending time outside in natural settings, (4) learning new things, and (5) offering time and care to others. The good news is that none of these five happiness enhancers requires consumption or money. In other words, none of them costs Earth anything![31]

In sum, reducing our footprints need not be framed in terms of deprivation and self-sacrifice. Instead, it can be seen as offering opportunities to experience genuine fulfillment by addressing our fundamental human need for relationship, connection, and meaning.[32]

Wrap-Up: Expanding Ecological Consciousness

> We have lived our lives by the assumption that what was good for us would be good for the world. We have been wrong. We must change our lives so that it will be possible to live by the contrary assumption, that what is good for the world will be good for us. . . . And that requires that we make the effort to know the world and learn what is good for it.
>
> —Wendell Berry[33]

The world that we create, both individually and collectively, will be determined, to a significant degree, by the quality and boldness of our questions. The question *Why*, as showcased in this chapter, is a particularly powerful tool for discovering the underlying causes of many troubling phenomena. For instance, every worker on a Toyota assembly line is taught to analyze problems by asking *Why* over and over:

This bolt fell off.

Why?

Because the thread is stripped.

Why?

Because it was misaligned with the screw.

Why? . . .

It turns out that almost every design problem encountered at Toyota can be solved with five or fewer *whys*.[34]

Fearlessly asking *Why* (as young children are prone to do) provides opportunities to be real with each other and honest with ourselves. For example, imagine a mother having this conversation with her ten-year-old daughter:

Daughter: I just learned in school that there are three billion people in the world who earn less than $2.50 per day[35] and that many of these people are hungry. Why is that happening?

Mother: Well, honey, it's just that adults in those places tend to have a lot of kids and there are just too many mouths to feed.

Daughter: But why do they have so many kids?

Mother: Well, dear, I don't know. Maybe they just don't know any better.

Daughter: But my teacher said that every day kids in those places die of sickness and hunger. Why is that?

Mother: [Pause.] Listen, hon, it's just that there is a lot of sickness in those places and sometimes there are bad droughts and not enough food.

Daughter: So, why doesn't anybody do anything to help?

Mother: Oh, they do. The Red Cross is there helping out in emergencies.

Daughter: Then why are all those children still dying?

Mother: I don't know; it's just the way it is. . . . Listen, it's not good for you to dwell on these things.

Daughter: Why not?

Mother: It will just make you sad.

Daughter: But why is it bad to feel sad . . . when that's what I feel?

Mother: Here, have a snack and stop this talk. There is nothing we can do.

Daughter: Why can't we give some of what we have to those children?

Mother: That would get really complicated. I wouldn't know where to start.

Daughter: Why wouldn't you know where to start?

Mother: Listen, you just need to accept that there is nothing we can do about this!

Having the courage to question limiting beliefs—for example, *There is nothing we can do about this*—creates space for the flowering of compassion and with this, the resolve to set things right.

Applications and Practices: Living the Questions

> I want to beg you, as much as I can, dear sir, to be patient toward all that is unsolved in your heart and try to love the questions themselves like locked rooms and like books that are written in a very foreign tongue. Do not now seek the answers, which cannot be given you because you would not be able to live them. And the point is to live everything. Live the questions now. Perhaps you will then gradually, without noticing it, live along some distant day into the answer.
>
> —Rainer Maria Rilke[36]

Almost everything we do as human beings is based on beliefs. We believe that it is necessary to eat three meals a day and so, without questioning, that is what we do. Ditto for driving cars, going to college, and wearing socks. These, and many similar practices, have become so completely kneaded into our psyches that we mistake them for essential truths. But what if we question these "essential truths"? What if we question everything?

Using Questions to Unmask Assumptions

Take a minute to carefully consider the following statement: *Technology will solve all of our problems.* Is this true? Do you agree? Why? How do you know? What if it's false? Read it again, this time with the awareness that it is simply an opinion—that is, an assumption that is worthy of scrutiny. Author and social critic Jerry Mander recommends that all new technologies be judged as "guilty until proven innocent" because, in his view, new technologies sometimes create more problems than they solve. By way of illustration, consider this vignette from Mander's book *In the Absence of the Sacred*:

> In the early 1900s the car was portrayed as a harbinger of personal freedom and democracy: private transportation that was fast, clean (no mud or manure), and independent. But what if the public . . . had been told that the car would bring with it the modern concrete city? Or that the car would contribute to cancer-causing air pollution, to noise, to solid waste problems, and to the . . . depletion of the world's resources? What if the public had been made aware that a nation of private car owners would require the . . . paving of the entire landscape, at public cost, so that eventually automobile sounds would be heard even in wilderness areas? What if it had been realized

that the private car would only be manufactured by a small number of giant corporations, leading to their acquiring tremendous economic and political power? That these corporations would create a new mode of mass production—the assembly line—which in turn would [contribute to] worker alienation, injury, drug abuse, and alcoholism? That these corporations might conspire to eliminate other means of popular transportation, including trains? That the automobile would facilitate suburban growth, and its impact on landscapes? What if there had been an appreciation of the psychological results of the privatization of travel and the modern experience of isolation? What if the public had been forewarned of the unprecedented need for oil that the private car would create? What if the world had known that, because of cars, horrible wars would be fought over oil supplies?

Would a public informed of these factors have decided to proceed with developing the private automobile? Would the public have thought it a good thing? . . . I really cannot guess whether a public so well informed and given a chance to vote, would have voted against cars. Perhaps not. But, as it was, the public was NOT so informed. There was never any vote, nor any real debate. And now, only [a few] generations later, we live in a world utterly made over to accommodate the demands and domination of one technology.[37]

Mander, in effect, is exhorting us to ask questions—inconvenient questions—questions that cause us to stop and really think about the seldom-acknowledged downsides that invariably accompany all new technologies. In this, I am struck by the strong connection in my life between car ownership and personal consumption. When I have a car in my driveway, I can (and sometimes do) respond to different shopping impulses; for example, *I think I'll just run out to the store now to buy x or y or z.* It is so easy; a car gives me ready access to a whole world of stuff, but I don't really *need* any of this stuff. How do I know? Because when a car isn't available, the urge to shop passes and I find that I get along just fine.

OK, now it's your turn. I invite you to use your critical thinking skills to ferret out common assumptions—that is, unquestioned beliefs—that crimp your own thinking. For example, maybe you believe that we, as a society, are making progress; or that our economy must continually grow if we are to flourish; or that competition is the best way to generate innovation; or that you, as a single individual, have no power to change things. Whatever assumptions you come up with, dig in and question each one with both fervor and courage.

Yes, courage is necessary. After all, questioning assumptions is not easy, especially when it involves opposing cultural norms. As author Derrick Jensen points out, our culture tends to shush us when we question its sacred icons:

"Question Christianity, damned heathen. Question capitalism, pinko liberal. Question democracy, ungrateful wretch. Question science, just plain stupid."[38]

When I first encountered Jensen's words, I confess to being taken aback. But, then I paused and wondered if capitalism really is the best economic system that we can aspire to. I also wondered about democracy—not its merits, but rather whether it's true that we, in the United States, live in a genuine democracy. Jensen's statement even prompted me to question my own profession—science—wondering about its inherent biases and limitations. All this leads me to conclude that questioning assumptions is a necessary first step for any substantive change, be it personal or cultural.

Using Questions to Distinguish *Wants* from *Needs*

The ability to distinguish wants from needs is important, insofar as we live in a consumer culture in which advertisers are intent on convincing us that those things that we may merely *want* are, actually, essential *needs.*

Here's an exercise that can be helpful in distinguishing your *wants* from your *needs.* All that's required is a pencil and paper and some curiosity. Begin by drawing a line, lengthwise, down the middle of the paper. In the left column (Column 1), list things that you regard as necessities—that is, the things that every human being needs in order to survive (e.g., air to breathe, food to eat, clothes to keep warm, etc.). Then, on the right side (Column 2), list twenty of your most important nonessential material possessions— that is, things you own and use (e.g., phone, toothpaste, bed) but that are not absolutely necessary for your survival.

Now, select five items from Column 2 that you would be able to give up for a while without too much difficulty. Put the letter *A* next to these five items. Next, select an additional ten items from Column 2 (bringing your total to fifteen) that you could give up but that to do so would cause you some inconvenience, along with some stress. Place the letter *B* next to these ten items. Finally, put a *C* next to the remaining five items in Column 2, because to give these up would cause you considerable inconvenience and genuine stress.

Now, with your completed table in hand, pick the *C* item that for you feels most like a necessity, even though—if you had to—you know you could survive without it. Once you have made your selection, resolve to live without this seemingly essential item for two days. If you aren't sure what to give up, consider placing your phone out of reach for two days. Granted, this won't be easy because many of us are addicted (to some degree) to the devices that we surround ourselves with, especially our phones.

Insofar as this might be true in your case, you may experience *withdrawal*—that is, pain in the form of upset, confusion, and imbalance. You can evaluate whether this is true by observing changes in your mental and emotional state. It will be important, as well, to pay attention to the ways that temporarily eliminating this one "C" item from your life either separates you from, or brings you into fuller relationship with, yourself and the world around you. If you have the courage to undertake this experiment, you may discover that the value of what you gain in self-knowledge will exceed any suffering that you might experience.

Using Questions to Make Wise Purchasing Decisions

Recently I asked the women in my environmental science class to raise their hands if they were wearing makeup. Most hands went up. I then asked, Why? I contextualized my question by saying that I found it interesting that adolescent girls apply makeup to try to look older, while older women often use makeup in an effort to look younger. What's that all about?

In response, a coed said, "It's like we're not OK, just as we are. Makeup advertisements have convinced me that I'm never OK unless I have makeup on." This woman was able to unveil the central strategy of most advertising—that is, to play on our insecurities, convincing us of our insufficiency, while promising that buying a certain product will bring us relief. Though this advertising approach sells lots of product, it leaves us (consumers) bereft of the one thing that could bring us genuine and lasting relief: the full-hearted acceptance of ourselves, just as we are.

We need not be victims. All that's required is that we pause to tap into our core values before deciding on a purchase. By way of example, imagine that you have just purchased your first home, and when the fall comes you notice that your lawn is covered with leaves. So you head out to the hardware store to buy a rake. Once there, you locate a sturdy, bamboo rake selling for $40. Should you buy it? As you stand there trying to decide, you notice that leaf blowers are on sale for only $59—not that much more than the price for bamboo rakes. You are tempted to go for the leaf blower. After all, you tend to like new gadgets, and this one will allow you to get your yard cleaned up in less time. But rather than rushing ahead with your purchase, you pause and ask yourself the question: "What will I be saying 'yes' to if I buy the rake instead of the leaf blower?" For some people, buying the rake would mean saying *yes* to quiet (instead of noise), *yes* to self-reliance (rather than machine reliance), *yes* to natural pace (instead of machine pace), *yes* to thrift (rather than unnecessary spending), *yes* to simplicity (instead of complexity), *yes* to vigorous exercise (rather than labor-saving devices), *yes* to clean air (instead of air pollution), and *yes* to durability (rather than built-in obsolescence). Of course for others, this discernment process could go in the direction of the leaf blower insofar as it promises speed, power, and convenience.

Questions for Reflection

- Over your lifetime, what has been your relationship with the question *Why*, and what's one *Why* question that arose for you while reading this chapter?
- What steps could you take to reduce your footprint size without compromising your well-being?
- What is a significant purchase that you made in the last three months? Looking back on that purchase, what can you infer about your values? In other words, in making your purchase, what were you saying *yes* to and what were you saying *no* to?
- As people learn about all of the environmental problems in the world, there is a tendency to throw up their hands and say that there is really nothing they can do. Why do you think people tend to respond in this way? How do you respond?
- When was the last time you identified one of your assumptions and then proceeded to question that assumption? What did you learn in the process? What's an assumption you are holding now that might be ripe for questioning?
- What is something important that you learned about yourself while reading this chapter?

PART III
Healing Ourselves, Healing Earth

There are hundreds of ways to kneel and kiss the ground.

—Jelaluddin Rumi[1]

The unequivocal diagnosis that emerges from part II is that Mother Earth is sick. Consider again some of the many signs of Earth's suffering: forest fragmentation; migratory bird declines; aquifer depletion; ailing seas; soil degradation; species extinction; climate destabilization; sea-level rise; chemical contamination of air, land, and sea; endocrine system disruption. These are not isolated problems; they are all connected to the explosive growth in human population and consumption and the concordant proliferation of waste that is being spewed into Earth's soil, water, and atmosphere.

Just as a human being who is experiencing stress slowly sickens, so it is that Earth now sickens, suffering under the weight of humankind's expanding ecological footprint. All of this is occurring, in no small part, because our old story—grounded in speciesism, domination, hyper-individualism, and incessant growth—is no longer working. Indeed, our times are tumultuous precisely because our old story is dying. Taking all of this into account, Eckhart Tolle, in his book *The Power of Now*, writes:

> At this moment in history, humans are a dangerously insane and very sick species. That's not a judgment. It's a fact. As a testament to insanity, humans killed a hundred million fellow humans in the twentieth century alone. No other species violates itself on such a grand scale. Only people who are in a deeply negative state, who feel very bad indeed, would create such a reality as a reflection of how they feel.[2]

Tolle, in effect, is suggesting that the world that humankind has created, pocked as it is with pollution, abuse, and strife, is a direct reflection—that is, a projection—of our collective inner world. Certainly a nerve is touched when we summon the courage to open our eyes and look straight-on at the suffering, separation, and violence now being inflicted, not just on humans, but on the entire family of life.

Allowing ourselves to literally feel the pain of our times is a necessary prelude to waking up. Already, throughout the world—in households, neighborhoods, businesses, churches, schools, in the countryside and in the city—an awakening is under way. Often it begins by asking simple questions:

- Is this it?
- Is this what it means to be alive, to be a fully expressed human being?
- Am I living the soulful life of purpose and meaning I was created to live?

The answer, most often is: "No, this is not why I am here." This "No" is important. Saying "No, I don't believe in this . . . No, I won't do this anymore . . ." is a prelude to creating a world that deserves our "Yes."

In the final two chapters of this book we describe the disabling stories and belief traps that have ensnared us, thereby leading to ecological havoc (chapter 7). This is followed by an elaboration, in chapter 8, of the *new story* of kinship and community and deeper purpose that is now emerging both within and among us.

7

The Old Story

Economism and Separation

> Those who do not have power over the story that dominates their lives, the power to retell it, rethink it, deconstruct it, joke about it, and change it as times change, truly are powerless, because they cannot think new thoughts.
>
> —Salman Rushdie[1]

Residents of southern India have a clever way of capturing monkeys. They drill a hole in a coconut, place rice inside, and then secure the coconut to a tree with a chain. Here is the clever part: The hole in the coconut is just large enough for a monkey's hand to slip inside, but if the monkey clutches onto the rice, his clenched fist will be too big to shake free of the coconut, and he will be captured by the villagers.

Just as our primate cousins are captured when they refuse to *let go* of the rice, we, too, can become trapped when we are unwilling to *let go* of counterproductive ways of thinking and acting.

An essential first step in addressing the environmental and social challenges of our times is to awaken to the fact that we live today—as has been true of humans down through the ages—in a culture-based story that significantly shapes our thoughts and our consequent behaviors. Most of us (and I include myself here) are not aware that we are playing parts in our culture's grand story. That's the way it is with cultural stories that are effective. We inhabit them so fully and seamlessly that they escape our notice.

The intention of this chapter is to explore how our culture shapes our beliefs—our story—regarding how we should live our lives. This examination wouldn't be necessary if our *story* was working, but there is a growing sense, in many quarters, that our story is coming to the end of its utility.[3]

WARM-UP: EXPLORING YOUR STORY OF SUCCESS

What does it mean to you to be successful? Find out right now by coming up with three words to complete the sentence: Successful people are: (1) _____; (2) _____; and (3) _____. If you are like many people in our culture, your success words will be associated with things like wealth, recognition, achievement, power, and status. But is this what you—deep down—really believe?

What would happen if you were to ignore your culture's teachings regarding success and instead define success on your own terms? By way of example, consider these words from Oberlin professor David Orr: "The planet does not need more 'successful' people, but it does desperately need more peacemakers, healers, restorers, storytellers and lovers of every kind. It needs people who live well in their places. It needs people of moral courage willing to join the fight to make the world habitable and humane."[2] How do Orr's words land in you?

Foundation 7.1: A Story Rooted in Separation?

Some years back one of my students suggested that I read Daniel Quinn's book *Ishmael*. Using a fictional format, Quinn explains how our fractured relationship with Earth is a product of Western culture's overarching story. Our story, in Quinn's view, is a living mythology that explains how things came to be and how we are to live our lives.

Quinn used Nazi Germany as an example of how a people's story can go awry. Hitler provided the German people with a story (narrative) that explained how the Aryan (white) race had been discriminated against and abused by mongrel races throughout history. His story went on to foretell how Germanic peoples would soon rise up and wreak vengeance on their oppressors, especially Semites, and in so doing, assume their rightful place as the master of all races.

Many Germans failed to see Hitler's story as misguided; instead, they embraced it as their destiny. Our times are not so different:

> Like the people of Nazi Germany, [we, too,] are the captives of a story. Of course, we don't even think that there is a story for us. This confusion exists simply because the story is so ingrained that we have ceased to recognize it as a story. Everyone knows it by heart by the time they are six or seven. Black and white, male and female, rich and poor, Christian and Jew, American and Russian . . . we all hear it incessantly because every medium of propaganda, every medium of education pours it out incessantly. . . . It is always there humming away in the background like a distant motor that never stops.[4]

In the United States today, we continue to live within a story. The story is transmitted to us in our homes, our schools, our churches, our daily news reports. An underlying theme of our story is humankind's inexorable *march of progress* and the special role of science and technology in contributing to America's exceptionalism. As proof of U.S. superiority, we are taught about the virtues of America's democracy, the power of America's military, the superiority of America's science, the wonders of America's agriculture, and so on.

One way to appreciate the power of a culture's story is to envision what it would be like to be born into a society with a very different story from our own. For example, imagine that your culture's story was grounded in the belief that you are inextricably connected to all that lives and that to be successful in life, your job is to manifest unconditional love and compassion everywhere, every day, toward everything. Can you sense the potential transformative power of this kind of story? Indeed, as futurist Barbara Hubbard points out: "As we see ourselves, so we become."[5] Herein lies the audacious power of a culture's story, for better or for worse.

Big Picture Story

The overarching story of Western civilization is predicated on the assumption that the world—Earth—was created for humans. Embedded in this story is the belief that conflict, including war, is the necessary means by which good triumphs over evil. No surprise, then, that we often frame life as a struggle or contest; for example, doctors *wage war* against cancer; couples *fight* for child custody; companies *campaign* for a greater share of the market; farmers *attack* pests; communities *combat* crime; economists *subdue* inflation; nations *battle* for resources . . . and on and on. In adopting these metaphors, our story conditions us to believe that winning, domination, and coming out on top are what most matter in life.

This *big-picture* story regards the appearance of *Homo sapiens* as a capstone event in the history of the cosmos. The Christian Bible teaches that creation was unfinished without man, as the world needed a ruler. Man was created to subdue the world. Still today, we hear it over and over: man is conquering the deserts; man is taming the oceans; man is controlling the atom and the human genome; man is gaining mastery over outer space. In sum, man is here to control, manage, and exercise dominion over Earth; this is man's destiny.[6] Implicit in this story is the belief that man is apart from and above the rest of creation. Moreover, the repeated use of the word "man" in our guiding story underscores the role of patriarchy in most human cultures today.

Of course, this notion that man is the climax, the final objective, of cosmic creation is simply a story that most of us have passively absorbed.

But really, it takes a fair measure of naïveté to believe that the entire cosmic unfolding of the universe, extending back some fourteen billion years (see chapter 1), was finally concluded when *Homo sapiens* appeared on a planet we happen to call "Earth." Lest we forget, since the appearance of man some two hundred thousand years ago, the universe has continued to expand; new stars have been born; and the principles and processes undergirding biological evolution and speciation have continued to operate, largely unabated, just as if man had never appeared. Declaring that man is the endpoint of Earth's evolutionary processes is no less shortsighted than imagining a period far back in time—say when photosynthetic bacteria were the most complex life form on Earth—and believing that the appearance of these amazing bacteria was the final objective of biological evolution. Clearly the generative creativity of the universe, much less that of the tiny "speck" that is Planet Earth, didn't stop with photosynthetic bacteria, nor has it stopped with us. We are minuscule players in a much, much grander unfolding.[7]

The Age of Separation

Today there is talk about how human beings are becoming more and more connected to each other via jet travel, the Internet, social networking, and so forth. But at the same time, it appears that we are becoming more separate from ourselves, from each other, and from Earth. This separation first shows up in childhood, when school kids are separated by age, test scores, and socioeconomic status; separation continues with adults as we submit to work routines that often distance us from realizing our life's deeper meaning and purpose; and it appears in retirement when we are isolated in special facilities, separated from the social dynamics of our communities.

Separation is pervasive, in no small part because those of us born in the West have inherited a way of thinking, called *dualism*, that conditions us to see reality in either/or terms—for example, black versus white, true versus false, superior versus inferior. Dualism conditions us to believe that humans are superior to nature and that my gender, my school, my values, my religion, my politics, my race, or my country is better than yours. Though dualistic thinking may be comforting, reality cannot be crisply parsed into discrete categories.

Take the example of good versus bad. As a child, I was socialized to believe that when I did something that others judged as *bad*, it meant that I *was* bad. Since then I have learned about the history of the aboriginal Orang Asli people of Malaysia. For them, the idea of labeling a person "bad" would be akin to punishing a plant that was not growing well. Historically, a person in their culture who was not behaving within the norms of the group was understood not as bad but as having forgotten his/her true nature. The solution was to bring that person more fully into the center of the community so that he/she would remember that he/she belonged to, and was an integral part of, something greater than himself/herself. Though we are not Orang Asli, we can, if we choose, step beyond assessments of *good versus bad* by extending compassion to those who have gone astray (including ourselves), rather than creating separation though acts of judgment and punishment.

The modern-day repercussions of our separation are evident everywhere. We abuse and even kill each other through acts of neglect, violence, and war while also heaping abuse on Earth's air, water, and land. Through the lens of psychology, it is possible to interpret the wounded world that we are creating as a manifestation of our own inner discontent, as evidenced by the growing prevalence of stress, anxiety, addiction, depression, and suicide in the United States and beyond.[8]

Economism: An Outgrowth of Separation Consciousness?

Imagine that you have been hired to make sense of human culture as it is enacted in the United States today. You set about your task, visiting homes, neighborhoods, shopping centers, churches, workplaces, schools, restaurants, factories, farms, sporting events, and entertainment venues across the country. Through your observations, you determine that many people, in particular those under sixty, spend the majority of their time working, with their free time devoted to shopping, screen gazing, and occasional socializing.

You note that these people receive money for their work, and they use their money to acquire all manner of stuff, much of which they discard within a year of purchase. You are also struck by the busyness of these people, by the amount of stress they seem to carry, and by their, predominantly, indoor lives.

Later, in trying to make sense of your observations, you encounter the concept of "economism." Like many other *isms* (e.g., socialism, Protestantism, nationalism), economism refers to a belief system. Proponents of economism see life through the lens of economic activity, for example, working, consuming, investing, inventing, manufacturing, marketing, and copywriting.

Citizens living within the story of economism believe that their primary purpose in life is to work hard to make money so that they will be able to buy the things that will bring them happiness. In effect, they live by the formula:

Work → Money → Possessions → Success → Happiness.

Economism has now become so fully integrated into the U.S. worldview that many of us simply see it as the way things are—end of story.[9]

A CRICKET AND SOME COINS

One afternoon, Gerry was walking along a busy sidewalk in Washington, DC, with a Native American friend from the Bureau of Indian Affairs. People were hustling along around them, and the sounds of honking horns and noisy car engines filled the air. In the middle of all the activity, Gerry's friend stopped and said, "Hey, listen, a cricket!"

"What?" said Gerry.

"Yeah, a cricket," said the Native American man as he peered under a bush and located the cricket.

"Damn," said Gerry. "How did you hear that cricket with all this noise and traffic?"

"It was the way I was raised," said the native man, "what I was taught to listen for." Then he reached into his pocket and pulling out a handful of coins—nickels, quarters, dimes—dropped them on to the sidewalk. As the coins hit the surface, all the pedestrians nearby turned their attention to the sound.[10]

What we listen for and what we hear reveal what matters most to us. Seduced by the story of economism, coins, not crickets—the world of commerce, not the world of nature—capture our attention.

A measure of economism's effectiveness lies in its ability to shape the purpose of many societal institutions and functions. For example, we have been taught to believe that the main purpose of schools is to teach children the skills they will need as productive, working adults so that U.S. businesses can be competitive in the global economy. In this same vein, we are taught that the main purpose of government is to promote policies that will ensure that the U.S. economy flourishes.[11] Under the rule of economism, the needs and values of business—the *growth economy*—have come to dominate society. Activities that generate a profit or bring high returns on investment are judged as desirable, often irrespective of whether they are wise, wholesome, or even morally defensible.[12] Meanwhile, endeavors that fail to generate large returns—no matter how beneficial or noble—are often marginalized.

Through the lens of economism, the primary purpose of Earth is to provide resources that will fuel the human economy. The commodification of Earth has become so routine that most of us barely register its oddness. But what if a multinational company attempted to commodify human bodies? After all, each of our bodies is composed of many different chemical elements, and some, like phosphorus, nitrogen, potassium, and iron, are valuable commodities. Though you might find the idea of commodifying human bodies—even dead ones—reprehensible, this is how our economy treats Earth's forests and grasslands and mountains and oceans and all the life that dwells therein—that is, as resources that we have a right to liquidate as we see fit.

In contrast to Earth's body, economism extends respect, even reverence, to the "almighty" dollar. I sometimes make this point with my students by holding a lit match in one hand and a crisp one dollar bill in the other. As I bring the match closer and closer to the dollar bill, I ask them to register what they are feeling—Excited? Afraid? Curious? Then, I proceed to set the dollar on fire. Some students react with a mix of confusion, shock, and anger to my action. They even say I should be arrested for such a heinous act, as if by burning a piece of paper I have blasphemed a sacred religious icon. After taking in their comments, I ask them to consider how burning that dollar bill might have actually benefited Earth. For example, might one less super-fluous purchase be a benefit to Earth, insofar as each product we consume has a measurable *footprint* on Earth's air, water, and land?

My larger motivation for this provocation is to challenge students to consider how money—especially organizing our lives around money— might separate us from becoming fully human in the very deepest sense. None of this is to suggest that I am in favor of abolishing money (more on this in the next chapter).

Economism as a Pseudo-Religion?

Economism, with its associated beliefs, has become so fully integrated into both U.S. and global culture that it now appears to act as a kind of pseudo-religion. Yes, metaphorically, economism has the equivalents of deities, ministers, missionaries, churches, commandments, and more. Though this metaphor might seem a bit extreme, I ask you to join me in playfully exploring it. Start by considering economism's *deities*—the things in our culture that we regard as all-knowing and all-powerful. Here are three candidates:

- The market—the omnipotent *father* that we are directed to trust in.
- Money—the primary means of earthly salvation that we spend our life energy pursuing.
- Technology—the source of knowledge and power that we are encouraged to embrace.

Indeed, these three (among other possible candidates) constitute a kind of *trinity* in which we place our faith. It follows that the *ministers* of economism—those preaching the messages of economism—include the likes of financiers, fiscal analysts, technologists, and politicians.

Meanwhile, the minions from transnational corporations, promoting neoliberalism and free-market capitalism, serve as economism's *missionaries,* while the outlets of the corporate-controlled media function as economism's *churches* as they spread economism's messages. And just like a bona fide religion, economism is also replete with commandments:

- Work long hours and you will be judged worthy!
- Invest in the market and you will be rewarded!

- Accumulate possessions and you will be happy!
- Trust in technology and you will be enlightened!

Upshot: The ideology of economism is now so deeply stitched into the American psyche that most people experience it as reality—that is, as the way life is and must be! For example, if you happen to be a college student, ask yourself: Why am I in college? What's my goal? What's the meaning and purpose of my life? If your answers center on the acquisition of money and security, you are an adherent of economism. Why? Because like other followers, you ascribe to the belief that your success and happiness in life will be determined, to a significant degree, by your ability to accumulate money. And yet this is a highly questionable assumption—one that was not generally entertained by earlier generations of college students. For example, in the 1960s, fully 86 percent of college freshmen registered their primary college concern as "developing a meaningful philosophy of life," whereas today only about 40 percent highlight this motivation.[13]

It would be easy to take this last statistic at face value and characterize today's college students as greedy and self-centered, but from where I sit, this judgment would be a mistake. In fact, when I invite students to reflect on what they most value in life, most refer to family, friends, religion, service, health, freedom, and adventure. By contrast, when I ask them to characterize the values of their culture, they frequently use words like *competition, consumption, speed, image, greed, convenience,* and *getting ahead.* I have been surprised by undertones of anger and cynicism in many respondents. Here are young Americans who have personal values that are, for the most part, generous and life-affirming, living in a culture that they perceive as crass. Though they may not name it as such, they seem to be experiencing a measure of cognitive dissonance resulting from the clash between their personal values and their perceived values of their culture.

This dissonance—think separation—was highlighted in "Yearning for Balance," a comprehensive analysis of American perspectives on modern life. The authors of this study concluded that when Americans—irrespective of gender, age, or race—look at the condition of the world today, they come to a similar conclusion: "Things are seriously out of whack. People describe a society at odds with itself and its own most important values. They see their fellow Americans growing increasingly atomized, selfish, and irresponsible; they worry that our society is losing its moral center."[14]

Foundation 7.2: The Price of Economism

Economism—with its formula: Work → Money → Possessions → Success → Happiness—has served us well in some respects, but it is not without many troubling consequences. Let's have a look.

Economism Separates Us from Our Common Sense

The story of economism tells us that if we hope to flourish, as individuals and as a society, we must keep growing, and growing, and growing. This constant growth translates to working more, exploiting more, and abusing more, all of which means consuming more. But does this system really make sense? Is it wise? What if we really don't need to keep growing in economic terms, but instead are now being called to grow in terms of creativity, interdependence, social justice, compassion, and wisdom?

Rather than a consumption/growth-based economy with its erratic fluctuations and wide disparities in wealth, what if we endeavored to create a *steady-state* economy—that is, an economy that meets people's needs without compromising Earth's life-support services?[15] Such an economy could bring us the genuine wholeness and well-being that we most truly seek—for example, whole bodies, healthy minds, heartfelt relationships, deep connections—all leading to a world of peace, health, and beauty. We've been conditioned to reject such propositions as impractical, but consider that what's truly impractical is continuing to give our life energy to stories, like economism, that wreak havoc on ourselves, each other, and Mother Earth.

Seriously, what if we don't really need to spend all our life energy working, rushing, and consuming? Specifically, what if working less could actually enhance human well-being? If you doubt that this is possible, take note that the average American worker today—as a result of constant technological innovations—is more than twice as productive, compared to the 1950s. This point bears repeating: An individual American worker produces twice as much *product* today compared to a worker in 1950. This means that, in theory, Americans could be working much less, given our doubled productivity, yet we are working more than ever!

Here is a thought experiment that points to what's possible. Begin by imagining that over the fifty-year period 2025 to 2075, U.S. productivity will double once again (a reasonable expectation given the current rate of technological innovation) and that U.S. workers could reap the benefits of this doubling in per capita productivity by having to work only about half as many hours (e.g., twenty to twenty-five hours/week instead of forty to fifty hours/week) while still receiving roughly the same salary.

Of course, our story barely allows us to imagine this. Instead, the logic of capitalism (the handmaiden of economism) dictates that employers must constantly strive to increase profits because it is primarily the accumulation of money that will ensure security and safety, and by extension, the well-being of employees. Caught in this story, companies have two possible responses to doubling worker productivity: (1) lay off half the workforce to reduce payroll costs while adding to shareholder profits, or (2) keep the workforce and increase profits by doubling production, thereby creating and marketing more stuff (much of which may actually undermine, rather than enhance, human well-being).

Remarkably, the idea of responding to a sharp rise in worker productivity by reducing work hours doesn't enter the rigid calculus of economism. But imagine if it did. Specifically, imagine how having more free time (because of the need to work fewer hours) could contribute to our well-being in genuine ways—for example, by giving us more time to take care of each other, more time to exercise citizenship; more time for joy, for play, for friendship, for creative expression; more time for self-care, for communing with nature; more time to volunteer, to express gratitude, to perform random acts of kindness, to become more fully and ecstatically human; more time to remember why we are here; more time to appreciate and care for Earth, our home.

Economism Separates Us from a Sustainable Future

What runs the global market economy? Is it money? Technology? Labor? These tools are certainly aspects, but the primary driver has been fossil fuels. If you are sitting inside right now, take a look around. Chances are that every object you see—this book, the walls, your chair, your clothes, your cell phone—was created using fossil fuels.

While burning fossil fuels allows us to produce the stuff that we want, it has also been causing potentially catastrophic climate warming (see chapter 5). Already, as of this writing, we have increased Earth's temperature by 1 degree Celsius, and that alone has been enough to melt more than half the summer ice in the Arctic; decimate enormous swaths of the world's corals; and unleash lethal wildfires, floods, droughts, and hurricanes worldwide.

Given this spreading chaos, researchers at Oil Change International, a Washington-based think tank, conducted a major study to determine the upper limit on the amount of fossil fuels in Earth's active coal, oil, and gas extraction operations that we can risk burning without cooking ourselves and the planet into oblivion. While their research required an enormous amount of number crunching, the conclusion can be expressed by comparing just two numbers. The first is 942 gigatons of CO_2, the combined amount of carbon dioxide ready for release in the world's *active* fossil fuel deposits. The second number, 353 gigatons of CO_2, earmarks the *upper limit* that the IPCC (Intergovernmental Panel on Climate Change) has established for the amount of carbon dioxide that we can release to the atmosphere while still having a break-even chance that Earth's temperature will not exceed the 1.5 degrees Celsius threshold that scientists now benchmark as the tipping point into climate chaos.[16]

Upshot: If governments, oil companies, and policy makers were guided by prudence and common sense, they would opt to leave the majority of

Earth's fossil fuel reserves in the ground, thereby avoiding the risk of generating even more climate chaos. But given economism's requirement for constant growth, it is hard to imagine that big energy producers will opt to keep their proven reserves below ground and out of circulation. After all, this stuff is already economically aboveground insofar as "it is figured into share prices and companies are borrowing money against it. These reserves are primary assets, the holdings that give companies their value."[17] As this book goes to press, fossil fuel entrepreneurs—both nations and corporations—continue to invest in fossil fuel exploration and extraction, even as they struggle to justify their actions.

In spite of all of these unsettling signs, there are hopeful reports, every day, about citizens and companies and governments mobilizing to forestall climate change by shifting to sustainable energy sources such as hydro, solar, geothermal, and wind; enacting *carbon footprint* reduction measures at state and local levels; and decarbonizing stock portfolios by shifting investments away from fossil fuels.

Economism Separates Us from Mother Earth

A fundamental precept of modern economics is that as individuals seek to maximize their self-interest, they will, without any conscious intention, benefit society as a whole. This comforting notion implies that we can do pretty much whatever we want without having to worry about the well-being of Earth.

To make this oversight more vivid, imagine that you are a CEO and your company owns, among many other assets, a hundred-thousand-acre expanse of mature forest. You acquired this land as an investment, and your shareholders expect a good return. Unsure of what to do, you consult with a professional forester. She tells you that if you cut your forest, it will take one hundred years to regrow to maturity. Given this time frame and the hundred-thousand-acre extent of your holdings, she advises you to harvest the trees from just one thousand acres each year. She assures you that by taking this approach you will realize a net profit of at least $1 million each year. You like the idea of maintaining a steady income stream by cutting just one thousand acres each year, all the while maintaining the health and diversity of your forest holdings. But before deciding, you consult with an investment analyst, who tells you that it would make a lot more sense, in economic terms, to clear-cut the entire one-hundred-thousand acre expanse of forest right now and then invest the hundred million in profit back into the market. He adds that even if you were to take a conservative approach and put your hundred million in stocks that only deliver a 3 percent return, you would still be guaranteed a three million dollar annual return—three times the measly one million promised to you by the forester. Yes, the forest will be gone, the soils degraded, and the wildlife partly exterminated, but your stockholders will be happy and you will be fulfilling your fiduciary responsibilities to them.[18] Economism is like this—it has its own logic that separates its adherents from considering the long-term well-being of Earth's ecosystems.

Economism Separates Us from Each Other

Economism, with its emphasis on work and busyness, can undermine what up to now has been essential to human flourishing—people living together in communities with their lives grounded in interdependence.

Until very recently, all human interactions were face-to-face. Today Americans spend, on average, eleven hours each day interacting with screens in the form of TV, smartphones, or computers, compared to a mere forty minutes per day engaged in face-to-face socializing.[19] The virtual world, while entertaining and stimulating, comes at a cost insofar as it separates us from the visceral experience of direct human contact.

When we are not preoccupied with our devices, we're clocking hours at work. The United States now leads the industrialized world in terms of the proportion of our population that is employed, as well as the number of days we spend working each year and the hours we work each day. Indeed, the average American's workload now exceeds that of Germans, French, and Italians by an average of 270 hours per year. Think of it as Americans working six and a half weeks more each year than many of our European counterparts.[20] Columnist Ellen Goodman captured this "new normal" when she wrote: "Normal is getting dressed in clothes that you buy for work, driving through traffic in a car that you are still paying for, in order to get to the job that you need so you can pay for the clothes, car and the house that you leave empty all day in order to afford to live in it."[21]

Just as our feverish devotion to work can undermine family and community bonds, the same can be said for the design of our living spaces. Let's start with house size. The floor space of the average U.S. home has ballooned from approximately 980 square feet in 1950 to approximately 2,600 square feet today.[22] Meanwhile, the average number of people per household has declined from 3.4 in 1950 to roughly 2.5 today. Translation: Each American now requires 1,000 square feet of living space; that's more than four 15-square-foot rooms per person—enough space to ensure that household members don't have to interact very much. Add to this the spread-out design of today's neighborhoods. For example, according to the U.S. census, in 1920 the average population density was ten people per acre for cities, towns, and suburbs taken together. Today, the population density is down to four people per acre, meaning that we are, on average, more than twice as spread out as we were a century ago, in spite of the fact that our population has more than doubled in the interim.[23]

Upshot: More space both within and between our homes, in conjunction with more time spent away from our homes, translates to fewer opportunities for community engagement. One consequence of this is what Bill McKibben calls "hyper-individualism"—the belief that we can meet all our needs on our own and therefore don't really need each other.

WALMART AND HYPER-INDIVIDUALISM

It has been documented many times: Walmart comes to town and the local economy—built upon community ties—begins to unravel. Ultimately, Walmart's success depends on the creation of hyper-individualistic societies that accord little value to community ties. As Bill McKibben observes, "For Wal-Mart to prosper, we must think of ourselves as individuals—must think that being individuals is the better deal. But, [if you] think of yourself as a member of a community . . . you'll get the better deal. You'll build a world with some hope of ecological stability and where the chances increase that you'll be happy."[24]

Ultimately, the *independence* that economism promises us is an illusion insofar as it makes us *dependent* on a vast, largely invisible, global supply network for food, clothing, fuel, and entertainment. Before we became entrapped in the ideology of economism, we had no need for this global network because we lived side-by-side and face-to-face, in communities built on interdependence.

Economism Separates Us from Abundance

Consider for a moment all the harmful things occurring on Earth today: the blasting away of mountaintops to get at coal, transforming magnificent tropical rain forests into cattle pastures, decimating populations of ocean fishes, toxifying Earth's air and water. Many are of the opinion that greed causes us to abuse Earth in these ways.[25] We hear it all the time: "People are just so darn greedy!" There's certainly some truth to this. What causes me or you or anyone to exhibit greed? Think about times in your life when you have acted out of greed. What was going on for you? Maybe it was as simple as taking a bit more than your fair share of ice cream?

I have observed that my greedy behaviors are invariably preceded by a fear that I am not going to have enough. For example, if I know that there is plenty of ice cream, greed does not arise. So, could it be that the problem isn't with greed, but with the belief—the *story*—that we live in a world of scarcity where there isn't enough to go around?

Scarcity can certainly be true in specific instances, but, in the big picture, we all live in a world of abundance.[26] For instance, even in the face of our burgeoning population and our poor stewardship of Earth's soils, there is still more than enough food being produced on Earth to adequately feed the entire human family. This same potential to meet everyone's needs also exists in the realms of housing, clean water, health care, and education.[27]

Although the capacity exists for everyone to be adequately provisioned, two billion–plus human beings—25 percent of humankind—live in abject poverty today. Meanwhile, just twenty-six multi-billionaires now possess a net worth that is equivalent to that of 3.8 billion human beings—half of the world's population.[28]

It is the perverse alchemy of greed, entitlement, and arrogance that drives the super-rich to accumulate more and more wealth. In the end, it seems that it's not wealth—not resources—that we lack, so much as humility, compassion, and generosity.[29]

Economism Separates Us from Happiness

The tragedy of economism is that this story isn't making us any happier. For example, though Americans have increased their expenditures by almost threefold since the 1950s, we are no happier today than in the 1950s—this according to the National Opinion Research Council. That's right; though we have more stuff than ever before—more education, more music, more electronics, more clothes, more entertainment, more cars, more spacious houses—we are no happier.

This is not to say that there is no connection between money and well-being. Researchers examining happiness in cash-poor countries like India, Brazil, and the Philippines confirm that an increase in per capita income from, say, $3,000 to $10,000 per year, results in a marked increase in happiness, but beyond $10,000—that is, beyond where basic needs are met—the correlation between income and happiness falls apart. This result is mirrored, anecdotally, in the fact that the happiness scores of *Forbes* magazine's richest Americans are identical to those of the Pennsylvania Amish—a people relatively poor in cash but rich in community bonds.[30]

Those pursuing more and more wealth often assume that because their first bump-up in income made them happier, having still more money will further increase their happiness. It's like having a beer and feeling good and then downing fifteen beers, believing that this will make you feel fifteen times happier. It doesn't work that way.[31] In fact, all the big stuff that big money buys—fancy cars, trophy houses—often leaves people feeling trapped; that is, rather than owning lots of cool stuff, their stuff, in a sense, ends up owning them.

One thing that *has* increased in the United States, in tandem with rising consumption, is the incidence of depression, now ten times higher than in the 1950s.[32] Psychologist David Myers refers to this pattern of rising wealth and diminished well-being as "The American Paradox." Specifically, Myers observes that at the outset of the twenty-first century, Americans find themselves "with big houses and broken homes, high incomes and low morale, secured rights and diminished civility. We [are] excelling at making a living but too often failing at making a life. We celebrate our prosperity but yearn for purpose. We cherish our freedoms but long for connection. In an age of plenty, we are feeling spiritual hunger. These facts of life lead us to a startling conclusion: becoming better off materially has not made us better off psychologically."[33]

Summing up, economism has many troubling impacts—both on individuals and on society—but in the end, economism is just a story. We suffer its negative effects only as much as we accept its version of reality.

Stepping Back to See the Big Picture

> You see everything is about belief; whatever we believe rules our existence, rules our life.
>
> —Don Miguel Ruiz[34]

Economism is grounded in the notion that Earth is primarily a lump of resources waiting to be transformed into products for our use. This worldview presents a depressing characterization of the human enterprise, suggesting that it is work, money, and the acquisition of material goods that brings meaning and purpose to our lives.

We can choose another story. Every day each of us creates stories as we try to make sense of the world around us. For example, say you are seated in a restaurant and someone you know walks in but fails to acknowledge you. How would you interpret this? That your friend didn't want to disturb you? That she is mad at you for some reason? That she forgot her glasses and didn't recognize you? Each interpretation of the situation is a story, even though—when you are caught in its grip—it will seem like the absolute truth. To underscore this point, consider this vignette from Byron Katie's book, *Loving What Is*:

> Once, as I walked into the ladies' room at a restaurant near my home, a woman came out of the single stall. We smiled at each other and as I closed the door she began to sing and wash her hands. What a lovely voice! I thought. Then, as I heard her leave, I noticed that the toilet seat was dripping wet. How could anyone be so rude? I thought. And how did she manage to pee all over the seat? Was she standing on it? Then it came to me that "she" was a man—a transvestite, singing falsetto in the women's restroom. It crossed my mind to go after him and let him know what a mess he'd made. As I cleaned the toilet seat, I thought about everything I'd say to him. Then I flushed the toilet. The water shot up out of the bowl and flooded the seat. And I just stood there laughing. In this case, the natural course of events was kind enough to expose my story before it went any further.[35]

It is easy to get tripped up in stories that create separation, as was the case with Byron Katie. When we are caught up in the story of economism, separation occurs as a result of our alienation from ourselves, from each other, and from Mother Earth. Handicapped in this way, we forfeit our birthright to become fully and wholeheartedly human. None of this separation and alienation is necessary. Just as Byron Katie awakened to her enfeebling story when she flushed the toilet and saw the water shoot up, we can awaken to the crippling effects of economism as we acknowledge its limitations and summon the courage to imagine a different story.

I sometimes offer my students an embodied experience of what it feels like to shift from an old, dysfunctional story to a new, enlivening one by placing a rope on the floor in the front of my classroom and then inviting five volunteers (the "A" group) to stand on one side of the rope, facing five volunteers (the "B" group) standing on the opposite side of the rope. When everyone is ready, I ask those in the "A" group to greet their partners with a feisty "Yo Mama!" I even have them practice their "Yo Mama" until they can say it with attitude. Then I ask those in the "B" group to respond with "Don't you *Yo Mama* me!"

When everyone is fired up, I tell them to reach out and grab their partner's right hand. Then, I explain that their task is to "do anything in your power to move your partner across the rope." They have only five seconds to do this. The "As" get ready by repeating "Yo Mama!" and the "Bs" respond with "Don't you *Yo Mama* me!" Then, I say "Get set, . . . Go!," and while the participants struggle to get their partners across the rope, I count down: 5 . . . 4 . . . 3 . . . 2 . . . 1. Usually, one of the two people in each pairing wins by dragging their partner to their side of the rope.

While they stand panting, I congratulate the winners. Then I ask both parties to engage in the activity a second time, but this time to approach it so that they both win. As they line up, I slowly repeat the instruction: "Your assignment is to do anything in your power to move your partner across the rope," and then I say "Go!" This time most pairs stand there dumbfounded, uncertain what to do, but usually there is at least one pair that merrily glides past each other, voluntarily exchanging positions. In this case, there are no overt winners or losers, just two people who have completed a task in a mutually satisfying manner.

Sometimes a student complains that I set them up to view this as a competition by goading them to say "Yo Mama," etc., and giving them only five seconds to "win." I acknowledge this, explaining that this is exactly what our culture does to us by conditioning us to see life as a competition, with winners and losers.

It need not be this way. For example, what would it take to see yourself and those around you with new eyes—for example, each as a kindred spirit rather than a competitor? This shift in perception could open the possibility for a new story, one centered on the question, "What can we do for each other?" rather than economism's soulless refrain, "What's in this for me?"

Wrap-Up: Expanding Ecological Consciousness

> Tell me the story of the river and the valley and the streams and woodlands and wetlands, of shellfish and finfish. Tell me a story. A story of where we are and how we got here and the characters and roles that we play. Tell me a story, a story that will be my story as well as the story of everyone and everything about me . . . a story that brings us together under the arc of the great blue sky in the day and the starry heavens at night.
>
> —Thomas Berry[36]

In this chapter we have entertained the possibility that Western culture's story—a story founded, to a significant degree, on conquest, control, dualism, hyper-individualism, consumption, scarcity, and separation—is at the root of today's socio-ecological malaise. In its modern guise of economism, this story proclaims: Humans are essentially economic beings, and the subjugation of nature is necessary for progress. We are all conditioned to believe in this manner—American, Russian, Chinese, Mexican—but this story is more alienating than bonding, more soulless than inspiring.

The solution is not to shun modernity. This would be folly, for as philosopher Ken Wilber reminds us:

> The rise of modernity . . . served many useful and extraordinary purposes. We might mention: the rise of democracy; the banishing of slavery . . . the widespread emergence of empirical sciences, including the systems sciences and ecological sciences; an increase in average life span of almost three decades . . . the move from ethnocentric to world-centric morality; and, in general, the undoing of dominator social hierarchies in numerous significant ways. These are extraordinary accomplishments.[37]

According to Wilbur, the time has come to take the good that modernity offers and leave behind its harmful effects, attitudes, and behaviors.[38] Certainly, a more life-affirming story is possible. For example, we could choose a story like this: We humans are born to connect, to learn, and to create in ways that respect and honor Earth, and it is in our very nature to seek to bring forth a world of cooperation, justice, beauty, love, abundance, and peace.

If we were to live within this life-affirming story, economism's soulless *trinity* of Market, Money, and Technology could be replaced by a *trinity* consisting of Community (the genuine source of security), Love (the force that ensures connection and well-being), and Spirit (the generative power that serves as the ground of existence). When we change our story, we will change our destiny.

Applications and Practices: The Power of Story

> Who would you be without your story? You never know until you inquire. There is no story that is you or that leads to you. Every story leads away from you. Turn it around; undo it. You are what exists before all stories. You are what remains when the story is understood.
>
> —Byron Katie[39]

Our most important story—the one that influences everything else in our lives—is the story we have about ourselves. Who we are? How did we get here? What's our purpose? Our answers to these and related questions comprise our personal belief system and worldview.

Because we tend to be seamlessly stitched into our worldview, we may struggle to actually see it. The movie *The Truman Show* is instructive in this regard. In this film, Jim Carrey plays the part of a man (Truman) whose entire life is a television show that is broadcast around the country. The remarkable thing is that Truman does not know this; he is just living his life, or so he thinks. The director of the television show, in effect, orchestrates Truman's life, determining when the sun will shine or clouds will appear and what will occur in Truman's life on any given day. During one episode, a reporter asks the director why it is that Truman is not able to figure out that his whole life is just a TV show. The director responds candidly, "We all accept reality as it is presented to us."[40] What about you? Are you consciously living your life, or has your life become mostly a habituated response to what is happening around you?

Discovering Your Worldview

Whether consciously articulated or not, our beliefs extend into every pocket of our existence, guiding our actions and decisions from birth until death. For example, if you have been conditioned to believe that birth should occur in a hospital, children should go to school, adults should get married and have kids, old people should retire, and dead people should be buried underground, then you will blithely follow this life script.

Without conscious awareness of our beliefs and how they shape our lives, we aren't nearly as free as we think. You can begin, right now, to identify the beliefs that comprise your worldview by completing the following open sentences:

1. _____ created the world.
2. The universe is _____.
3. Earth is here to _____.
4. Earth's plants and animals are here to _____.
5. Human beings' purpose on Earth is to _____.
6. The good and bad things that happen on Earth are the consequence of _____.

See your responses to these prompts as puzzle pieces comprising parts of your worldview—that is, the story that determines, at least in part, the purpose, direction, and meaning of your life.

If you want to explore how new ways of seeing might open you to new possibilities, note your responses to these six open sentences a second time, but this time, respond in such a way that you create an entirely different story. So, for example, if, in the first round, you wrote, "*God created the world*," in your second round you could conjure another answer to this question. The idea—and this is the important part—is to make your second set of responses more original and mind-expanding than your first set. It is not necessary that you believe your second responses, just that you exercise your innate creativity and fecund imagination to come up with new possibilities that—though they may seem a bit far-fetched—are fresh and have the potential to expand your consciousness.

If you want to go deeper still, perhaps give thought to these four questions: (1) How does your second *story/worldview* leave you feeling? (2) What would happen if you began to act and live as if this new story were true for you? (3) What would you lose? (4) What would you gain?[41]

Discovering Stories in Your Daily Life

The Inuit people have a saying that the Great Spirit must have loved stories, for why else would he have created so many people? Indeed, each of our lives is a story, with each day as a chapter.

We can adopt simple practices to awaken to the stories in our lives. For example, you could regard all your daily encounters and activities as having a beginning, middle, and end; or see each meal as a story involving food preparation, eating, and cleaning up; or view the evening news as a good-night story; and so forth. To formalize this practice, consider doing an "evening review" of the events of your day. In other words, run the movie of your day through your mind, paying particular attention to the activities in which you engaged, the interactions you had with others, and the thoughts and deliberations surrounding any choices you made during your day. Think of this evening review as a way of harvesting the stories of your life.

STORY HARVESTING

I was waiting to board an airplane. The line was long and moving slowly. Standing in a state of impatience, I watched a flight attendant guide an elderly couple onto the plane. They were both large people, and it appeared as if the man had suffered a stroke because his speech was slurred and his posture bent. A few minutes later, when I approached my seat at the back of the plane, I noticed that the same couple was seated in my row, and I remember feeling disgruntled by the possibility that I would have to squeeze in next to them, but as it turned out, my seat was on the other side of the aisle.

Later, about halfway through the flight, a flight attendant came down the aisle holding up a $20 bill, asking for change. It occurred to me that I might have change, but I decided that it would be too big of a hassle to undo my seatbelt and dig out my wallet to conduct a search. Just after I reached my decision, I heard the large man with the garbled speech say that he thought he had change. For several minutes, he wrestled with his seatbelt and then his wallet. Finally, he produced change for the twenty and the flight attendant thanked him.

It wasn't until that evening, when I paused to review the events of my day that I was able to fully consider the significance of this event, recalling first my chagrin when I thought that I would have to squeeze in with the couple and then a feeling of humiliation because it was that bent-over man who had helped the flight attendant when it would have been so much easier for me to have done so. I was humbled by his kindness; he was my teacher both then and today.

Author Greg Levoy recommends that we pause from time to time to embrace our life stories as grand myths because, in his words: "[Myths] get at the heart of human behavior, at profound truths, universal truths, ageless patterns. They are . . . stories of transformation: from chaos to form, sleep to awakening, woundedness to wholeness, folly to wisdom, from being lost to finding our way."[42]

In the end, each of us has a story, uniquely our own, to enact. It is easy to forget our power and become subdued by our culture's insistent pleas:

- Look out for yourself!
- Avoid unnecessary risks!
- Control your emotions!
- Conform!

These are the messages generated by the soul-crushing story of economism. To the extent that we acquiesce—sacrificing our souls for security—we risk having the lights of spirit and consciousness grow dim within us.

Questions for Reflection

- What are the stories that you were told when you were growing up, and how have these stories shaped your understanding of both yourself and the world?
- Of all the topics broached in this chapter, which one was most difficult for you to understand and accept? Why?
- How did you react to the story of burning a dollar bill? Insofar as this is a teaching story, what did it teach you about yourself?
- Do you live in a world of *scarcity* or one of *abundance*? How do you know? How does your story about this affect your life's meaning and purpose?
- In what ways are you separate from yourself . . . from other people . . . from Mother Earth? What might you do to heal separation in these three realms?
- If *economism* is Western culture's present story, what is an alternative story, waiting to be born?

8

Birthing a New Story
The Great Turning

> It was an old story that was no longer true. . . . Truth can go out
> of stories, you know. What was true becomes meaningless, even a
> lie, because the truth has gone into another story.
>
> —Ursula K. Le Guin[1]

In her book *In the Tiger's Mouth*, Katrina Shields shares a story about a young Irish woman who was working as a cashier in a Dublin supermarket in the early 1990s. The woman had read about the apartheid regime in South Africa—where the white minority was subjecting the black majority to grossly unjust political, legal, and economic discrimination; later, when she learned that South African blacks were asking those in the developed world to boycott South African produce, she decided that she would honor their call. This meant that when customers came to her checkout counter, she rang up everything except for the apartheid-tainted produce.

What happened next? Well, customers complained to the store manager about this clerk's refusal to ring up South African produce and, in short order, the woman was fired. But when the other cashiers heard about this, they joined the woman's protest by also refusing to sell the South African produce. Soon clerks from other Dublin markets also stood up in support of the woman's stance. When the media jumped on the story, the woman got her job back, and her supermarket stopped carrying South African produce.[2]

Each of us, like that woman from Dublin, can choose to stand up and say "no" to old stories—that is, old ways of seeing—that create separation rather than connection, war rather than peace, fear rather than love.

Just as we, as individuals, can choose to let go of old stories that leave us feeling fearful and submissive, the same is true for society as a whole. Indeed, humanity is now being challenged, just as people were in the time of Galileo and Copernicus, to adopt a radically new worldview. Pain and bewilderment (as well as excitement) surface anytime a person, or a people, experience the dissolution of their old story, and it is no different in our time. No single one of us knows with any certainty what lies ahead, but

---

LOVE RATHER THAN FEAR

When I turned fifty-five, my partner, Dana (forty at the time), expressed her deep desire to become a mother. At first I rejected her suggestion, believing that I was too old, and that it was a mistake to repeat life phases, and that people (including my own grown children) would judge me as foolhardy and irresponsible. But then I began to investigate my story, asking questions like: Who says I am too old? Who says it's a mistake to repeat life phases? As I did this, I was able to create a new story—one not contaminated with the fear of how I might be judged by others, but instead, one stabilized in trust and love.

---

many concur that we live in a fecund time that is pregnant with chaos and danger, as well as with possibility.

In this final chapter, we invite you to join minds and hearts with hundreds of millions of others around the world in the birthing of a new story. How? By looking and listening for the signs of The Great Turning—that is, humankind's *turning away* from crippling separation and *turning toward* each other on the path to becoming fully and ecstatically human.

Foundation 8.1: Seeing with New Eyes

Pick any contemporary environmental or social issue: climate change, ocean acidification, deforestation, endocrine disruption, species extinction, war, sea-level rise, genocide, racism, sexism—and in every case, separation—our *turning away* from each other—lies at its root. But we are not doomed, at least not yet. The Great Turning, emerging all around us, challenges us to see ourselves, each other, and Earth with new eyes—the eyes of interdependence, compassion, and love.

Seeing Mother Earth with New Eyes

In the previous chapter, I posited that, from the perspective of economism, Earth is an object from which we extract things wherever and whenever we please. This objectification of Earth often shows up in how we use language. For example, in English, we routinely spell "earth" with a lowercase "e" preceded with the article *the*, as in "the earth is the third planet out from the sun." Saying "the earth" implies that Earth is separate from us—that we are here and "the earth" is there. As Unitarian minister Michael Dowd points out, "To refer to the literal ground of our being, the source and substance of our life, as 'the earth' is to objectify. . . . Such objectification encourages us to continue seeing Earth merely as a resource for human consumption."[3] In contrast, when we use our planet's proper name, "Earth" with a capital E, we honor Earth's integrity as a creative, self-organizing, living presence that animates life in all its forms.

Take this a step further, now, by considering the question: Do you live *on* Earth or do you live *in* Earth? Which is it? Most people respond that they live *on* Earth because saying "I live *in* Earth" sounds strange. But the truth is that we actually live *in* Earth! If you doubt this, go outside and lie on your back and look up at the sky. In particular, observe the clouds, with the full knowledge that Earth is spinning in an easterly direction at more than a thousand miles per hour. So, if this is true, why aren't those clouds up there zooming westward across the sky at a thousand miles per hour?[4] This doesn't occur because all that stuff up there that we blithely refer to as "sky" or "atmosphere" is a part of Earth. It's all one! For confirmation, simply hold your breath. In so doing, you will be reminded that, just as fish are utterly dependent on water for air, our support medium is the atmosphere.[5] As a way of knitting this into human consciousness, author David Abram recommends changing the spelling of Earth to E*air*th.[6]

Taking all this in, it is easy to see that the expression "Mother Earth" is an accurate description for what Earth is to us: she is the womb of our very lives as she was for our ancestors and will be for our children and their children to come. Richard Nelson captures it this way:

> There is nothing in me that is not Earth, no split instant of separateness, no particle that disunites me from my surroundings. I am no less than Earth itself. The rivers run through my veins, the winds blow in and out with my breath, the soil makes my flesh, the sun's heat smolders inside me. . . . The life of the Earth is my own life. My eyes are the Earth gazing at itself.[7]

I find Nelson's imagery—Earth's winds in our breath, Earth's waters in our veins—to be insightful, yet heartbreaking, especially when I acknowledge that Earth's winds and waters—and by extension, my own breath and blood—are now laced with toxins as a result of our human hubris, ignorance, and alienation (see chapter 5).

If we are to survive as a species, we will need to awaken to the fact that Earth *is* our larger body. This means that when you walk along a sidewalk or forest path, you *are*, quite literally, Earth *walking*. Yes, each of us is a physical manifestation of Earth. This realization, when experienced in our bellies and bones, allows us to see and experience ourselves as part of Earth's body, our every cell an integrated, pulsating constituent of Earth. The same is true for all of Earth's creatures: bee, beaver, beetle . . . all woven together into Earth's tapestry of life.

Lauck's story (see box on the next page) suggests that in extending kind regard to a fly we might actually become more peaceful and loving toward each other. It appears that the Dalai Lama agrees, because when he was asked what he thought was the most important thing to teach our children,

---

SEEING INSECTS WITH NEW EYES

Joanne Lauck, in her book *The Voice of the Infinite in the Small*, tells a story about Joanna Macy, a teacher/activist who traveled to a village in Tibet to help people there start a craft cooperative. One day Macy met with a group of Buddhist monks. During the meeting, it happened that a fly fell into her tea. Though this was not a matter of concern for Macy, the young monk sitting beside her leaned over and asked if she was OK. Macy responded that she was fine. After all, it was just a fly. A few minutes later the monk expressed his concern once more but, again, Macy assured him that she was OK and he need not worry. Hearing this, the young man gently lifted the fly out of Macy's tea and left the room.

Out of sight, he carefully placed the tea-soaked fly on a bush until s/he revived. Returning to the room a few minutes later, the monk beamed as he reported to Macy that the fly was going to be OK. This was a poignant moment for Macy, because she saw that the young man's compassion extended not just to her but also to the fly, the lowly fly!

In interpreting this story, Lauck writes: "Our sense of self expands when we extend our compassion to insects. It is an appropriate response to our interdependence. The question is not how to connect with a fly—we are already connected. The question is how to translate that connection into appropriate behavior. Helping one in need is always a good place to start."[8]

---

he responded, "Teach them to love the insects!" His wise counsel is timely, especially as many children these days have been conditioned by our culture to fear insects—seeing them as both annoying and scary. But just think: If all of us—adults and children alike—could learn to appreciate and respect insects, in all their wondrous and mysterious forms, we just might be able to do the same or better when it comes to respecting and loving each other!

Seeing Each Other with New Eyes

Modern history illustrates that though humans are certainly capable of kindness and love, we frequently fail to exhibit these qualities. Yet, recent research shows that we are more attuned to each other than previously imagined. For example, Daniel Goleman, in his book *Social Intelligence*, reports that scientists have discovered that whenever we see a photograph of someone displaying a strong emotion, whether it be happiness, grief, rage, or confusion, the muscles of our face automatically begin to mirror the emotion we are observing, and these shifts in facial muscles can induce shifts in our feeling state.[9]

The response is neurological. For example, if I were to prick your finger with a pin, neurons would light up in a certain part of your brain. This reaction is to be expected, but what has astonished researchers is that if you, as a third party, watched as I pricked another person's finger, *your* brain would also light up in the same places as the brain of the person who was being pricked. This occurs because the human brain is wired with a class of cells dubbed "mirror neurons." These neurons fire in response to the emotions that the people around us are displaying, in effect, grafting their emotional life onto ours.

Goleman puts it this way: "Mirror neurons ensure that the moment someone sees an emotion expressed on your face, they will at once sense that same feeling within themselves. We are connected by our emotions as they are experienced not merely by ourselves in isolation but also by those around us—both covertly and openly."[10]

We are, it seems, wired for empathy, or as Marc Barasch puts it, "I feel you in me."[11] In sum, mirror neurons "suggest that an archaic kind of sociality, one which does not distinguish between self and other, is woven into the primate brain."[12] It is as if the Golden Rule is, in some measure, inscribed into our very biology.

Recent research also suggests that we are born predisposed to cooperate. For example, Michael Tomasello, in his book *Why We Cooperate*, reports that if children less than two years old see that an adult (even one they have just met) is unable to open a door because his hands are full, they will rush to his assistance. In a similar vein, if an adult pretends to lose an object, a one-year-old will point her to it, or if the adult drops an object in front of a young child, the child will instinctively pick it up and give it to the adult without being asked. Tomasello makes the case that such cooperative/empathic behaviors are not taught or performed for a reward; rather, young children are cooperative by nature.[13] However, by the time kids reach the age of three their natural predisposition for cooperation may be thwarted, especially if they live in a culture that encourages individualism and the separation that this promotes.

In some indigenous cultures, a sense of belonging and interconnectedness shapes all human relations. For example, Martin Prechtel, a Native American who lived for many years among the Tzutujil people in Guatemala, relates that in his village if a family had a sick child, the family, as a whole, understood itself to be sick. Likewise, if several people in the village were sick, the entire village experienced itself as sick, and this same consciousness extended to the land surrounding the village. For them, to imagine that they could be healthy if the village or the surrounding forest was suffering from disease would be as absurd as you or me saying, "I've got a fatal liver disease, but that's just my liver—I'm really fine!" Just as our sense of self includes our liver, for the Tzutujil, "self" extends to the entire community of life that they dwell within.[14]

Accepting rather than rejecting a human being whom we perceive as a threat is a departure from the predominant storyline of human history.

THE POWER OF EMPATHY

You don't have to be a Tzutujil native from Guatemala to experience the Other as a kindred spirit. Anyone we feel estranged from can become our teacher. Case in point: Saint Louis native Angie O'Gorman was awakened late one night by an intruder standing over her bed. Somehow, amid her fear, O'Gorman knew that she and the intruder were both human—that is, connected to each other—and for her this meant that they would both be damaged by what ensued, or they would both emerge safely, with their integrity intact. This insight allowed O'Gorman to act with empathy for both herself and the intruder.

She began by speaking calmly into the darkness, asking the man what time it was.

"Two-thirty," he replied.

She expressed concern that his watch might be broken because the clock on her nightstand read 2:45. Then, after a pause, she asked how he had entered her home. He said he broke a window. She said this was a problem for her because she didn't have the money to fix it. He confessed that he was also having money problems.

When she felt it was safe, O'Gorman told him, respectfully, that he would have to leave. He said he didn't want to, had no place to go. Lacking the force to make him leave and seeing a person without a home, O'Gorman said that she would give him a set of sheets, but he would have to make his own bed downstairs. The man went downstairs, and O'Gorman sat up in bed for the rest of the night. The next morning, they had breakfast together. Then the man left.

By turning toward this intruder with empathy instead of anger or fear, O'Gorman was authoring a new story—one in which no one wins unless everyone wins because if someone loses, everyone loses.[15]

Indeed, much of human history revolves around stories of who killed whom—that is, tales of aggression and vengeance. But in our time, "redemptive violence" is finally being seen for what it is: a misguided attempt to avoid experiencing our own inner pain and brokenness. As theologian Richard Rohr explains, "If you don't transform your pain, you will always transmit it. . . . You will always find someone who is worthy of your hatred . . . and then, of course, you become the evil you despise."[16]

This moment in history is challenging us to shift from a consciousness based in *exclusivity* to one rooted in *inclusivity*. Inclusivity is grounded in relationship, whereas exclusivity is rooted in separation. A consciousness centered in inclusivity generates trust; one moored in exclusivity breeds fear—especially, fear of the *Other*. When our goal is exclusivity, we silence those with whom we disagree; when inclusivity becomes our goal, we strive to create a world that works for all.[17]

Seeing Ourselves with New Eyes

In the end, how we see ourselves affects everything. If we believe that we are fundamentally separate from each other and from Earth, we will live in ways that continue to engender separation.

Certainly our culture—insofar as it is grounded in competition, control, and individualism—predisposes us to experience ourselves as separate from each other and from Earth. For example, we are conditioned from an early age to think of ourselves in terms of our names, our beliefs, our possessions, and so forth. But if you stop to think about it, you may discover that you are really not your possessions, not your job title, not your physical appearance, not your emotions, not even your thoughts.

You might be thinking, "How could I *not* be these things?" For example, "How could it be that I am not my thoughts?" Well, for one thing, if you are like the rest of us, most of your thoughts are actually opinions and thought patterns that you have inherited from the world around you—for example, from your parents, your teachers, your friends, the media. When we identify with our thoughts and mistake them for who we are, we experience confusion. After all, our thoughts change from minute to minute; there is nothing permanent about them.

What about other aspects of your identity? Is your name who you are? Clearly not, because you can change your name and you will still be you. What about your religion? If you drop your old religion and get a new one, will you have a new "I?" No, it will be the same "I," just with some new attitudes, insights, and beliefs tacked on. So, go ahead and change your clothes, your car, your college, your job, your body, your country—change all of these things, and your essential nature will not change. Indeed, it is curious that most of the things that we say that begin with the words *I, me,* or *mine* refer only to surface attributes—having little to do with the essence of who or what we are. Upshot: We can't truly know ourselves until we break free from socially conditioned ways of seeing ourselves.[18]

The disorienting consequences of misperceiving ourselves are illustrated in a story told by Anthony DeMello. In the story, a man finds an egg of a wild bird and places it in the nest of a barnyard hen. Eventually the egg hatches and, low and behold, out comes a baby eagle. The eaglet is reared with a brood of chickens and conforms to their behaviors, scratching the ground for worms and insects, clucking and cackling, and thrashing her wings in feeble attempts to fly. Then, one day when she was is two years old, she looks up and is awestruck by a magnificent bird flying high in the sky. When she asks a nearby chicken, "What bird is that?," the chicken responds, "That's Eagle. She is different from us; she belongs to the sky. You are a Chicken; you belong on the ground." So it was that the barnyard eagle suffered from a case of "mistaken identity" until the day that she died.

Might it be much the same for us—that is, that our culture has conditioned us to see ourselves as *less than* we really are, flawed, never enough, when in fact we are each infused with goodness, compassion, and unique sensitivities—each capable of things beyond our wildest imaginings?[19]

How is it for you? Do you appreciate and accept yourself, just as you are? Are you enough? As a means of exploring this, I sometimes invite my students to pair up and simply gaze into each other's eyes without speaking. Many find this too challenging to even attempt; others try, but after a short time, break down in nervous laughter. But there are always a few who are able to calmly gaze into their partner's eyes, demonstrating that it is possible.

Many adults also struggle when asked to do this activity. This prompts the question: Why do so many of us find it difficult to calmly gaze into the eyes of a fellow human being? Doesn't this seem a bit strange? Are we afraid of what we might see? Or perhaps afraid of being seen just as we are—in a state of repose, without words or facial machinations? Might our struggle with this task also say something important about how we have been conditioned by our culture to see ourselves as never OK, never good enough, never quite worthy of appreciation and acceptance, just as we are?

Foundation 8.2: A New Story

The story of the United States—its formation and history—though pocked with darkness in places, has had an overall ascendant quality. Is this still the case today? Rather than an *exceptional people*, might it be that we have become a *troubled people*, who are now in search of a new story—one with the power to awaken us to what it means to be fully and resplendently human?

The Great Turning narrative, as mentioned at the beginning of this chapter, captures both the tumult and possibility of our times.[20] This story invites us to see ourselves: not as fighters on the battlefield of life but as collaborators, dwelling in a miraculous universe; not as selfish and brutish sinners, but as innately kind and compassionate souls; not as drones meant to spend our lives with our noses to the grindstone, but as beings on an exciting journey to ever-deeper awareness and ever-expanding consciousness.

Cultures do change. Things that once seemed normal—for example, greed, stress, alienation, numbness, anxiety—can become untenable if people choose to boldly explore new ways of being. For example, I recently took my students on a forest hike. Knowing that we would be out for a couple of hours, I brought along snacks consisting of nuts, grapes, apple slices, and crackers. At a rest stop on our walk, I told my students to find a stick about two feet long. Then I gave everyone two sturdy rubber bands with the instruction to fasten the stick to their right arm, thereby ensuring that their right elbow wouldn't bend. Only then did I set the snacks out on a picnic table, inviting everyone to help themselves, with the proviso that they hold their left arm behind their back, using only their right arm (the one held straight by the stick) to partake in this mini-feast.

Handicapped in this way, some tossed bits of food into the air, hoping that something would land in their gaping mouths; others sheepishly put their lips to the table in an attempt to vacuum up food bits. None of this was easy or pleasant.

As their patience began to fray, I suggested that they could transform their frustration into happiness if only they could create a *new story* for how they might partake in the snacks. That's all it took. Soon they were using their straight arms to feed each other, taking delight in the food and in the act of nourishing one another.

The Purpose of an Economy Is to Serve People

A successful *economy*, in its simplest formulation, is a network of relationships created to help ensure that everyone within a community is provided for. But this ideal is hardly the case today, considering that the richest 1 percent of humans beings now possess more wealth than the rest of humankind combined.[21] In light of this grotesque inequality, economist David Korten has called for a shift toward "a life-serving economy," in which the goal is to generate a good living for everybody rather than vast wealth for a few.[22] In such an economy, Earth's soils would be respected (rather than abused), her waters and atmosphere would be protected (rather than polluted), and her climate would be safeguarded (rather than jeopardized).[23] The centerpiece of such a *life-serving economy* is no different than the centerpiece of life itself: relationship, connection, community!

Genuine communities possess a unique form of wealth known as "social capital." People generate social capital when they trust one another, share talents and resources with each other, join with each other on community projects, and stand by and for one another in times of hardship. So it is that communities rich in social capital are able to transcend, to a significant degree, hardships imposed by limited access to financial capital. In this sense, social capital acts as an internal insurance policy, giving people something real that they can count on—each other—in hard times. Research reveals that people living in communities rich in social capital tend to be healthier and happier—for example, they are better able to cope with trauma and resist illness and are less likely to suffer from depression.[24]

It Takes Time—Not Money—to Create Social Capital

Time is money! Although this is a common expression, it's not true. *Time is life!* We live in time; we don't live in money.[25] In fact, our fixation with money can lead us to experience time as scarce.[26] And what does extra cash matter if you are engaged in work that fails to enliven you?

Aware of this disconnect, increasing numbers of people today are opting for more life rather than more money. This shift was first noted in the 1990s, when many Americans reported that—if given the choice between a pay raise versus more free time—they would choose more free time. And this wasn't just empty talk. For example, Boston College professor Juliet Schor reported that in the first decade of the present century, millions of Americans were actually reducing the time they devoted to work, either by cutting back on hours or working fewer jobs. Roughly half of these so-called *downshifters* cited job stress as the reason for pulling back; the rest were seeking more balance/meaning in their lives.[27]

Whatever the rationale, the big dividend for downshifters is more free time. Specifically, time to spend with loved ones, time to learn new skills, time to devote to dormant passions, and time to help friends and neighbors in need, not to mention time for worthwhile household activities like childcare, cooking, home repairs, and growing food—tasks that downshifters previously used their hard-earned income to pay others to do. In this vein, Schor points out:

> All activities, whether they are monetized or not, have the potential to yield returns. We recognize wages and salaries as the returns on employment. But activities that do not earn dollars create returns as well. Doing work in one's own household, without a wage, is production. The cooked meal, the completed tax return and the cared-for child all have economic value. . . . Even spending time with a friend, something one might consider purely uneconomic, strengthens social networks of support and reciprocity.[28]

Downshifters have, in effect, a different understanding of what makes for a rich life. For them, it's generally not about a new car or a bigger house or more possessions. Rather than exhausting themselves working to acquire nonessential stuff, they opt for fewer hours so that they can have more time to devote to what matters most in life: social bonds, joy, health, acts of creation, play, caregiving. In the process, downshifters create social capital while extending care to Mother Earth by reducing their *ecological footprint* (see Chapter 6).

Downshifting doesn't appear to be a passing fad. Schor reports that even those downshifters who occasionally missed having more income reported that they were still happier than before.[29]

Social Capital in Action

Wherever we live, the potential exists to create social capital. It all begins as we *turn toward each other* to ask: What can we do, with the resources available right here where we live, to make our community healthy and whole? Specifically, how can we join our heads, hands, and hearts to meet our food, energy, and shelter needs in ways that are sustainable and life-affirming? Let's have a look.

Food: As recently as a hundred years ago, much of the food consumed in the United States came from the fields and pastures surrounding our communities. Today, if you buy your food at a supermarket, it comes to you via the *industrial food system*—an impersonal approach to food production that involves enormous land parcels, mechanized planting and harvesting, the copious application of fertilizers and pesticides, extensive processing, and lots and lots of transporting. In fact, in the United States, the average food molecule travels fifteen hundred miles before reaching our plates.[30]

This industrialization of our food system requires huge amounts of energy. So huge that about ten times more energy (think fossil fuels) is expended in the production of the food that we eat than this food actually contains. Translation: If your lunch today supplied you with a thousand calories, it is likely that roughly ten thousand calories of energy were used to grow, process, package, transport, and prepare that lunch.

Many communities are taking steps to rectify this upside-down energy equation. Progress comes in the form of local, organic, small-scale farming operations—the opposite of today's industrial food-production model. It was in the 1990s that Americans began to signal their disillusionment with industrial food, in tandem with their growing enthusiasm for local, fresh, nonprocessed foods. Farmers responded by creating a socioeconomic farming model dubbed community-supported agriculture (CSA).

A typical CSA is composed of community members, from several dozen households, that pledge to support a local farm. Each household makes a payment at the start of the growing season in return for a share of the farm harvest. In cases where a community member can't afford to purchase a farm share, the option exists to work for several hours each week in exchange for a food share.

CSA families (shareholders) receive fresh produce, on a weekly basis, throughout the growing season. Early in the season, their weekly food boxes might be laden with lettuce, spinach, baby beets, turnips, radishes, scallions, and snap peas; a bit later—in June—the boxes will contain strawberries, chard, broccoli, zucchini, kale, green beans, and herbs; then in July and August, carrots, tomatoes, sweet corn, cucumbers, peppers, and raspberries make their appearance. Finally, toward the end of the growing season, shareholders receive bounteous offerings of onions, squash, cabbage, parsnips, potatoes, and pumpkins. Locally produced eggs, honey, breads, and meats also figure into many CSA operations.

CSAs are great for community members because participants receive local produce on a weekly basis, and CSAs make sense ecologically because the food is produced using much less energy than is the case with the far-flung, industrial food model. Moreover, CSAs are good for farmers because the growers receive, up front, the cash they need to finance their farming operations.[31]

This community-based approach for the production of food and the generation of social capital is catching on. The first CSA in the United States was established in the Northeast in 1984; today there are over 12,000 CSAs—an average of more than 250 per state.[32]

Along with the rise in CSAs, farmer's markets are also flourishing. Their number, currently in excess of eight thousand in the United States, has increased almost fivefold since the mid-1990s, making these open-air markets the fastest-growing component of the U.S. food economy.[33] In contrast to supermarkets, where most people act as anonymous consumers, farmer's market patrons have the experience of being *participants in a community* as they delight in the colors, smells, and tastes of fresh food while conversing with their neighbors, including those who grew the food that they will have for dinner.

Energy: Just as with food, communities are also seeking ways to meet their energy needs locally. Based on studies from the Institute for Local Self-Reliance, most of America's fifty states could meet all or most of their energy requirements using renewable energy sources located within their borders. For example, wind turbines and rooftop solar panels could supply more than four-fifths of the power needs for the state of New York. In other parts of the country, the local energy mix would differ—for example, hydropower resources predominating in the Northwest, solar in the Southwest, and biofuels in the Southeast.[34]

At present, there are more than nine hundred citizen-based, not-for-profit energy cooperatives in the United States, with more emerging each year.[35] In this new paradigm, energy, instead of being a commodity that people purchase from someplace far away, becomes a product that cities and towns, and even individual citizens, produce and distribute to each other. For example, if you are a homeowner, it may already be possible to work with an energy cooperative in your community to supply all your electricity needs by installing solar panels or small wind turbines on your roof, and it could even turn out that you would have electricity left over to share with neighbors.[36]

Shelter: For much of our human history, we lived in tight-knit groups, joining together to meet our shared needs for food, energy, shelter, and companionship. Things are different now, as most people, especially those living in "developed" countries, choose to live independently—physically separate—from each other (see chapter 7). Concurrent with this trend is a growing incidence in loneliness. For example, in a 2018 U.S. nationwide survey, 56 percent of twenty thousand respondents reported that they "sometimes or always feel that no one knows them very well," and 40 percent said that they "lack meaningful relationships and companionship" and feel "isolated from others."[37]

So-called ecovillages are a timely response to the worldwide fragmentation of communities. At present, the Global Ecovillage Network connects an estimated ten thousand communities around the world.[38] What these ecovillage communities have in common is an intention to significantly reduce their ecological footprints by sharing space, time, resources, creativity, and goodwill. Ecovillage members typically co-own things such as tools, appliances, play spaces, guest rooms, laundry facilities, workshops, and vehicles. Members also frequently join together for the growing of food and for routine tasks such as maintenance, cooking, and childcare.

Ecovillage structures are usually created by the community members themselves, with an emphasis on simplicity and beauty, using design principles that maximize energy efficiency and minimize waste. Dwellings are not spread out, as is typical with suburbia, but clustered together, linked by paths and walkways, with vehicle parking relegated to the village periphery. These layouts result in a significant reduction in ecological footprints, along with strengthened community bonds.

If you find the idea of living in an ecovillage appealing but can't imagine creating one from scratch, you could choose to simply see the neighborhood where you now live as an ecovillage-in-waiting—that is, as a place where you could join with neighbors to meet each other's needs. In fact, this very idea is being promoted by the Transition Towns movement that started in England in 2005. Today, so-called Transition Towns are present in fifty countries, with more than 150 Transition towns and cities present in the United States alone (e.g., Los Angeles, Charlotte, Chicago, Pittsburgh, Milwaukee, Fort Collins, Tulsa).[39]

Transition Town participants recognize that the time has come to turn *away* from our current growth-based fossil-fuel economy and *toward* a life-serving, sustainable, steady-state economy grounded in a deep respect for the living Earth that sustains and nurtures us. Such a transition calls us to reimagine our entire way of life in the realms of food, energy, housing, education, governance, and more. Think of it as an economic, cultural, and spiritual renaissance aimed not at *independence* (which is actually a myth) but at *interdependence* (the underlying reality of existence). In view of this, Transition Town participants acknowledge: "*Transition* is a social experiment on a massive scale. What we are convinced of is this: If we wait for the governments, it'll be too late; if we act as individuals, it'll be too little; but if we act as communities, it might just be enough, just in time."[40]

Creating a Collaborative Economy: Opportunities for Participation

A quiet revolution is brewing as people endeavor to create new economic models that genuinely enhance and enrich community life. Transportation is one more realm where change is happening. Many of us simply assume that car ownership is a necessity for modern life. Certainly cars can be very helpful at times, but owning a car is expensive and maintaining one can be a hassle. So, what if you could have the benefits of a car without the bother of owning one? Enter Zipcar, a community-oriented car-sharing company with a million members, operating in nine countries and five hundred cities.

Imagine yourself as a Zipcar member, freed from the hassle of car ownership. It's Tuesday morning, and you need a car. No problem. You simply go to the Zipcar website (or tap the Zipcar app on your phone) and reserve one of the Zipcars stationed in your neighborhood. Done! By not having to bother with car purchase and upkeep, Zipcar members save up to $600 per month in car-related expenses.

Zipcar also makes ecological sense because cars, like many other things that Americans own, are seldom in actual use; for example, the average privately owned car is on the road for only an hour a day. Upshot: Every Zipcar on the road replaces seven to eight privately owned vehicles because by becoming car-sharers, many Zipcar members no longer require private cars.[41]

And it's not just cars. We purchase many other products that we use infrequently. For example, an estimated fifty million Americans own a power drill, but the average electric drill is used for less than an hour during its entire lifetime; the rest of the time it sits idle.[42] Clearly, it is not necessary for everybody on the block to own their own power drill, much less their own hedge pruner, ladder, mower, snow blower, wheelbarrow, and so forth. So what if, just like a library for the sharing of books, there were neighborhood tool libraries? Actually, this idea is not a "what if" because it's already happening in cities around the United States, including Columbus, Atlanta, Berkeley, and Philadelphia. Tool libraries are an acknowledgment that we don't have to own tools; all we need is occasional access to the services that tools provide.[43]

Peer-to-peer renting is another strategy employed in collaborative economies. Sites such as Turo, Zilok, and Rentoid now act as platforms allowing citizens to loan things—such as their personal cars, tools, and electronic equipment—directly to each other. In a related development, citizens are even sidestepping the banking industry to extend cash loans directly to one another. In fact, 10 percent of all loans in the United States are now made peer-to-peer. This alternative makes sense insofar as it circumvents "asset managers, mortgage brokers, pension and mutual funds advisors and the big banks themselves, who for the most part [have] introduced faceless transactions and overhead, while removing the community loyalty that [is] the glue in person-to-person lending."[44]

Freecycle is a final illustration of a community-based initiative that models nature's ethic of mutual exchange. Grounded in the belief that "what's mine is yours," Freecycle operates in ninety-five countries, creating conditions (via the Internet) for people to supply each other's needs. Each day, worldwide, more than twenty-four thousand items, weighing approximately seven hundred tons, are exchanged.[45] In times when we have more than we need, we can act as Freecycle givers, and in times of need, we can become Freecycle receivers.

From an environmental perspective, these community-based exchange networks reduce both consumption and waste, because when things are shared, there is less need to manufacture new things. In fact, for every pound of new product manufactured, whether it be a smartphone or a sofa, approximately thirty pounds of waste are produced.[46] This means that if you used Freecycle to give your old sofa to someone else, not only would you be keeping approximately one hundred pounds (the weight of the sofa) out of a landfill, but you would also be eliminating the waste that would be generated in the construction of a new sofa.[47]

In their own ways, each of these sharing initiatives is the result of people seeing themselves not as independent from each other, but instead as interdependent members of a community.

Stepping Back to See the Big Picture

> Strange is our situation here upon Earth. Each of us comes for a short visit, not knowing why, yet sometimes seeming to [have] a divine purpose. From the standpoint of daily life, however, there is one thing we do know: that we are here for the sake of others.
>
> —Albert Einstein[48]

Humans are capable of behaving in selfish and brutish ways, but deep down, knitted into our biology, is a predisposition to care for one another. Don't take my word for it. Simply call to mind something you have done in the last week that made life better for someone else. Now, recall how you felt at that time. If you are like the overwhelming majority of us, you probably felt pretty good.[49]

Could it be that our predisposition to help one another comes from our innate gratitude for the gift of life that has been bestowed upon us? You didn't do anything to earn your life. In those first days, months, and years, there was someone there, giving you the gift of care. This is no trivial matter. "Imagine finding yourself plunged into an alien world in which you were completely helpless, unable to feed or clothe yourself, unable even to distinguish where your body ends and the world begins. Then, kind beings come and hold you, feed you, take care of you, love you. . . . Wouldn't you feel grateful?"[50]

Still, today, in the very moment that you are reading this, you continue to be the recipient of gifts. The air that you are breathing is free for the taking—a gift. The same is true for water: you didn't manufacture it; it is here, in abundance, a gift. It's no different with the food that you eat or the seeds that grow your food or the wood that created your house. Earth offers these things to each of us as gifts. We didn't create Mother Earth, much less ourselves, as is highlighted in these words from Satish Kumar:

> We do not own our intellect, our creativity, or our skills. We have received them as a gift and grace. We pass them on as a gift and grace; it is like a river which keeps flowing. All the tributaries make the river great. We are [each one of us] a tributary that adds to the great river of time and culture, the river of humanity. If a tributary stops flowing into the river, if it becomes individualistic and egotistical, if it puts terms and conditions before it joins the river, it will dry and the great river will dry too. To keep the great river flowing, all tributaries have to join in with joy and without conditions. . . . We need not hold back, we need not block the flow. This is unconditional union. This is how society and civilization are replenished.[51]

As we come to recognize that we are each recipients of gifts beyond measure, it is natural to feel gratitude, and with this gratitude comes an innate desire to reciprocate.[52]

A Gift Economy

Our guiding story—economism—often encourages separation, and with this separation comes environmental and social breakdown, but economism is not the only story open to us. We can, if we choose, devise other transaction systems to meet our daily needs. For example, some people, down through the ages, have chosen to cocreate *gift economies*, grounded in the perception that we live in a world of abundance, not one of scarcity. To the extent that this is true, there is no need to hoard. For example, if you have twenty ears of corn today and you only need two ears, why not give the other eighteen away? Then the recipients of your corn—your gift—will experience gratitude and quite naturally want to reciprocate. Think of it this way: through giving, we in effect create an empty space—a vacuum—that tugs upon the whole, ensuring that gifts flow back to us. Charles Eisenstein, in his book *Sacred Economics*, explained it this way:

> Receiving and giving go hand in hand. . . . To give and to receive, to owe and be owed, to depend on others and be depended on—this is being

fully alive. To neither give nor receive, but to pay for everything; to never depend on anyone, but to be financially independent, to not be bound to a community or place, but to be mobile. . . . [S]uch is the illusory paradise of the discrete and separate self. Independence is a delusion. The truth is, has always been, and always will be that we are utterly and hopelessly dependent on each other and on nature.[53]

Though this idea of a gift economy may sound like pie-in-the-sky idealism, it appears to be one of the ways that humans are configured to behave and respond. For example, the free online encyclopedia Wikipedia is the result of tens of thousands of people giving their time to create billions of articles of information that are now available, for free, to all of us. Meanwhile, the Internet, as a whole, also allows us to act as both givers and receivers, as we exchange news, videos, music insights, and much more. None of this is to deny the utility and effectiveness of money as a means of exchange in some circumstances, but by limiting ourselves exclusively to monetary transactions, we miss out on opportunities to cultivate the interdependence, joy, and gratitude that characterize gift economies.

We Are Here to Discover and Give Our Gift(s)

When you were growing up, did anyone ever ask you what your gift was—that is, what you were born to give to the world? If not, I ask you now, what is your gift? What is the beautiful thing that you are here to give? It need not be grandiose; it just needs to be yours to give. In this vein, Hawaiian poet and community organizer Puanani Burgess tells of a time when she asked a group of high school students to each tell three stories: the story of their *name*, the story of their *community*, and the story of their *gift*. All was going well until Burgess came to a certain young man. He did fine with the stories of his name and community, but when it came to telling the story of his gift, he bristled: "What, Miss? What kind gift you think I get, eh? I stay in this Special-Ed class and I get a hard time read and I cannot do that math. And why you make me shame for, ask me that kind question, 'What kind gift I have?' If I had gift, you think I be here?"

In that moment, Burgess felt awful for shaming the young man. Then, two weeks later, she spotted him down one of the aisles in the local grocery store. Still feeling ashamed, she tried to flee, but he caught up with her and said: "You know, I've been thinking, thinking, thinking. I cannot do that math stuff and I cannot read so good, but Aunty, when I stay in the ocean, I can call the fish and the fish he come, every time. Every time I can put food on my family table. Every time. And sometimes when I stay in the ocean and the Shark he come, and he look at me and I look at him and I tell him, 'Uncle, I not going take plenty fish. I just going to take one, two fish, just for my family. All the rest I leave for you.' And so the Shark he say, 'Oh, you cool, brother.' And I tell the Shark, 'Uncle, you cool.' And the Shark, he go his way and I go my way."[54]

This is a story of a young man with a remarkable gift—namely, the awareness and sensitivity to be in intimate relationship with the more-than-human world.

Wrap-Up: Expanding Ecological Consciousness

> I don't know, of course, what's going to happen, but it seems to me, imaginable, that a time could come when we will either have to achieve community or die, learn to love one another or die. We're rapidly coming to the time, I think, when the great centralized powers are not going to be able to do for us what we need to have done. Community will start again when people begin to do necessary things for each other.
>
> —Wendell Berry[55]

We could look at all the change, upheaval, and disturbance occurring throughout the world today and conclude that things are falling apart and that the *human project* is doomed. However, in the natural world, major breakthroughs and transformations are often preceded by great turmoil, even chaos. For example, Nobel Prize winner Ilya Prigogine demonstrated that certain chemical systems actually shift to greater order after they are disturbed. In other words, disorder can actually act as an ally, cajoling a system to self-organize toward a more coherent state, better attuned to the demands of its changed environment.[56]

If this seems counterintuitive, it might help to consider it from the perspective of your own life, considering that times of personal hardship and suffering can sometimes lead to personal growth and increased integrity. Social systems work in the same way: disruption and chaos are often necessary to awaken creativity and galvanize resolve. Seen in this light, the very unraveling of our old worldview—a view based on separation from nature, economic exploitation of Earth's body, and dominance relations among people—may be providing the creativity and energy required for the birthing of a new story. As Margaret Wheatley, author of *Leadership and the New Science*, observes:

> In the dream of dominion over all nature, we believed we could eliminate chaos from life. We believed there were straight lines to the top. If we set a goal or claimed a vision, we would get there, never looking back, never forced to descend into confusion or despair. These beliefs led us far from life, far from the processes by which newness is created. And it is only now, as modern life grows ever more turbulent and control slips away, that we

are willing again to contemplate [that the] destruction created by chaos is necessary for the creation of anything new.[57]

A new worldview won't take hold because of clever arguments or impassioned pleas; it will emerge only after many, many people come to understand that our present worldview/story is causing irrevocable damage and must be abandoned. To paraphrase Einstein, a new story begins to gain traction when people come to understand that the problems they face can no longer be resolved from within the same level of consciousness that created those problems in the first place.

Applications and Practices: Awakening

> See yourself moving more slowly through life, taking time to notice things, time to find your balance as you walk, time to notice how things look and smell around you. See yourself looking deeply into other people's eyes as you talk with them, studying their faces with attention and sensitivity. See yourself deeply enjoying the pleasure of love-making, a fresh salad, a starry evening sky, walking barefoot in wet grass. Imagine yourself gazing steadily inward, knowing and accepting yourself, your feelings, longings, spiritual intuitions, dreams.
>
> —Marc Burch[58]

If we are to survive the convergence of crises that now envelop us, it will be because we summoned the courage to *turn toward* each other, and in so doing, gained the capacity to see each other with the wide-open eyes of interdependence, not the shackled eyes of separation. There is hope. Indeed, these days people all over the world are awakening as they seek to discover what it means to be fully human.

Awakening by Turning Inward

Mindfulness meditation is a particularly powerful awakening tool. Variations of this practice are now being taught to people in all walks of life—business personnel, teachers, professional athletes, students—for the purpose of increasing concentration, awareness, insight, and overall well-being. In its most basic form, mindfulness meditation consists of sitting quietly while placing attention on one's breathing: simply noticing each inbreath and each outbreath. Although it may sound easy, this can be quite challenging, especially at first. For example, often our minds will introduce a new thought before we have even completed one inbreath, or if we do manage to complete a breathing cycle, our minds will jump in to congratulate us.

If you doubt this, pause now and give mindfulness meditation a try. Begin by assuming an alert but relaxed sitting position. Then cast your eyes down and direct your attention to your breathing—inbreath . . . outbreath—simply witnessing your inhalations and exhalations.

As you engage in this simple practice, you will probably discover that your mind is a very active place. The value of meditation is that it allows us to bring awareness to the mind's chatter. With this awareness comes understanding and—given sufficient practice—a semblance of freedom from our conditioned ways of reacting, seeing, and being.

You will know that you are making progress when you are able to assume the role of an observer, witnessing your thoughts as they come and go, without getting caught up in them. The idea with meditation is to welcome and acknowledge each new thought and then to release it. For example, if you get caught up in worry, simply note "worry mind" on your inbreath and then let go of *worry mind* on your outbreath. Do the same for "planning mind," "judging mind," "fearful mind," and so forth. The simple act of calmly observing and then naming our passing mind states brings us back to the present moment—the only moment there is.

Normally, we engage with our passing thoughts and get caught up in their emotional charge—that is, we *become* our thoughts. "In our so-called normal state of consciousness, we are, therefore, continually lost in the dramas of our lives, unaware of how the process that creates our stories is taking place."[59]

Meditation teacher Wes Nisker uses a movie theater analogy to explain how mindfulness meditation works:

> When we are watching the (movie) screen, we are absorbed in the momentum of the story, our thoughts and emotions manipulated by the images we are seeing. But if just for a moment we were to turn around and look toward the back of the theater at the projector, we would see how these images are being produced. We would recognize that what we are lost in is nothing more than flickering beams of light. Although we might be able to turn back and lose ourselves once again in the movie, its power over us would be diminished. The illusion-maker has been seen.[60]

It's the same with mindfulness meditation. As we look into and explore the workings of our minds, we become more present, more conscious, more awake, more free.

Awakening by Turning Outward

Just as turning inward is important for personal awakening, turning outward, toward each other, is essential to the shared quest to become fully human. Opportunities for *turning outward* abound in the neighborhoods and communities where we live.

1. A Neighborhood Gift Circle

Gift circles are a great way to build collaborative economies by fomenting interdependence. This community building is easy to do. Just gather with some neighbors; ten is a good number. Then invite everyone to take a turn sharing a need that they have. It could be anything. For example, maybe you, yourself, need someone with a strong back to help you put in a garden bed or to build a stone wall; or perhaps you need someone with technical skills to help you solve a computer problem; or maybe you would like someone to lend you an electric drill or a pasta maker or a ladder. After each person speaks, others in the circle can, if they choose, offer to meet the need or provide ideas for how the need could be met.

Then comes a second round, in which each person has a turn to tell about something they have to give. It could be something that they no longer need, like a toaster oven or a chair or art supplies; alternatively, it could be a willingness to lend tools or to offer assistance with bike repair. Here, too, as participants offer their gifts, anyone can simply say, "I would like that" or give the name of a person or organization that might benefit from the gift being offered.

Gift circles conclude, quite naturally, with expressions of gratitude both for the tangible expression of needs and gifts and the gift of each other's company.

The power of gift circles is that they break the human economy down to its essentials: needs, gifts, and connections. Regular gift circle participants discover that they don't need to spend as much time working to earn money because some of their needs are now being met by those in their gift circle. Once a culture of gifting takes root, it has a way of building on itself, because *what goes around comes around.*

2. What's Possible Here?

As Jay Walljasper of the Project for Public Spaces points out, "The neighborhood is the basic unit of human civilization. Unlike cities, counties, wards, townships, enterprise zones, and other artificial entities, the neighborhood is easily recognized as a real place. It's the spot on Earth we call home."[61]

So, imagine if each of us invited those in our neighborhood to a potluck centered on the question, *What's possible here?* By having a potluck, we would already be showcasing some of the things that are possible—for example, hospitality, good conversation, creativity, kindness, laughter, music, play, storytelling. Then, over dessert, everyone could share one wild idea for *what's possible here* by simply completing the open sentence, WHAT IF. For example, *what if* we removed the fences separating our properties and refashioned them into benches for sitting and swings for swaying? Or *what if* we envisioned our neighborhood intersections as gathering places with bulletin boards, tables and chairs, free-cycle bins, and food stands? Or *what if* we encouraged bartering by exchanging homemade bread and homegrown

---

READY, GET SET, GO!

Here's a recipe for promoting neighborhood kinship. The objective is simple: create a gift for your neighborhood. There are four guidelines to ensure success. First, the gift has to be useful. Second, it has to be handmade and created together with others. Third, it can't cost any money; that is, the necessary materials and tools have to be accessed without the exchange of money. Fourth, it has to be accessible to people of all ages.[62]

---

vegetables? Or *what if* we started growing u-pick strawberries and blueberries along our sidewalks? And while we are at it, *what if* we joined together to make a map of our neighborhood, including an inventory of each household's gifts and talents, as well as its needs? Yes, there are a thousand ways to rekindle our humanity, and most of them are possible right where we live.

In sum, when it comes to creating community, it's not necessary that we have a grand plan; it's enough to simply begin—to take that first step. Once we begin, we can be sure that forces, both within and outside ourselves, will come to our aid to guide us forward.

Questions for Reflection

- What *new story* about your life are you ready to announce? What is one step you could take today to begin to birth this new story?
- Imagine that starting tomorrow, you have four extra hours each day to use in any way you would like. What are two new enlivening things you would consider adding to your life? Now imagine that starting tomorrow, you have four fewer hours each day. What are some deadening things that you would eliminate from your life? Finally, take note that if you choose to remove these deadening things, you will be creating time and space to include the two enlivening things that you came up with!
- What yearly income do you think you need in order to be content? Now, imagine that your income will be half your preferred amount. How would you survive? How will you eat; how will you get around; how will you shelter yourself, etc.? Can you think of ways that your life could actually become better, more interesting, freer, or more creative, with your income cut in half?
- How might your conception and use of money change if you choose to believe that wealth results from how much we give, not how much we take?
- What's a personal gift you possess that you have not fully acknowledged? What are the consequences of holding your gift in exile?
- What's a "door" in your life that you are afraid to open? What's your story about what's on the other side of that door? What's a new story that would allow you to push open that door?

Epilogue

Mis estimados: Do not lose heart. We were made for these times.

—Clarissa Pinkola Estes[1]

Some years ago on a tranquil Sunday afternoon, I heard a high-pitched shriek coming from outside. My six-year-old daughter Katie was playing in the yard with her friend, Olivia. I rushed to the back porch and called out: "Is everything OK?!?" Breathless, Katie exclaimed, "A butterfly; we just saw a monarch butterfly!" Ahhh, to see with fresh eyes the wonder of a butterfly!

Biologist Elisabet Sahtouris has suggested that the process of metamorphosis, whereby an engorged caterpillar transforms into a delicate butterfly, is an appropriate metaphor for what is now happening on Planet Earth. Think about it: Caterpillars are prodigious consumers, able to eat many times their body weight in a single day and, in the process, to sometimes destroy their host plants. It seems that we humans, given our own prodigious consumption, have been acting like voracious caterpillars, gobbling up everything in sight.

There is more to this comparison. When a caterpillar is so stuffed that she can hardly move, she attaches herself to a leaf or a branch and forms a hard shell around her body. Then, most of the cells of the caterpillar's body break down, forming a kind of soup, but some cells, located in pockets in the caterpillar's skin, retain their integrity. Remarkably, these surviving caterpillar cells contain, within their DNA, the instructions for creating a butterfly. Because these cells can "imagine" a new way of being and living, biologists refer to them as "imaginal cells."

In this time of crisis, we humans need a radical transformation along the lines of what monarch caterpillars achieve. Like the monarchs, humankind has the equivalent of *imaginal* cells in the form of individuals who hold an *image*—a vision—for an entirely new way of living, one that is graceful, generous, gentle, and genuine. Indeed, after transforming into a butterfly, monarchs serve the world, acting as pollinators—giving back to life. It's time for us to do the same.

In this vein, I draw hope from the story of the poet Robert Desnos, a prisoner in a Nazi concentration camp during World War II. One day Desnos, along with a group of other prisoners, was ordered onto a flatbed truck. Desnos, and probably most of the others, knew that the truck was headed to the gas chamber, but no one spoke. When they arrived, the guards ordered

179

them off the truck. As they began to shuffle toward the gas chamber, Desnos grabbed the hand of the woman in front of him and, leaning in close, began to read her palm. The forecast was good: a long life, many grandchildren, abundant joy. Hearing this, a person nearby offered Desnos his palm. Again, Desnos foresaw a long life filled with happiness and good fortune. Other prisoners became animated, eagerly thrusting their palms toward Desnos, and in each case, he foresaw a long and prosperous life. The guards became visibly disoriented. Minutes before, they were confident in their mission, the outcome of which seemed inevitable, but now they were tentative in their movements. In fact, Desnos was so effective in creating a new reality—a new story—that the guards eventually ordered the prisoners back onto the truck and transported them back to their barracks.[2]

In this story, Desnos acted as a kind of imaginal cell—someone with the potential to imagine and then create a different reality. What about you? Can you conjure the world that you long for? Give this a go right now by completing the open sentence, *I want a world where* . . . When I invite my students to do this, I hear things like:

- I want a world where there are no weapons, no wars, no hatred.
- I want a world where children are cherished and nurtured and where elders are honored.
- I want a world where everybody is *family* and where people share rather than hoard.

Yes, when given the opportunity to speak, many of us express our longing for a kinder, gentler world—a world infused with care and compassion and honesty.

What is the world that you long for? What would be eliminated? What would be introduced? Can you paint it? Can you describe how two strangers would greet each other? Place your hand on your forehead—the locus of cognition—and describe the world that you long for. Then, when you are ready, hold your hand over your heart—the seat of compassion—and ask: *What is the world that I long for?* Do you have the courage to imagine it, the will to speak of it, the determination to play your part in creating it?

We are living at the time of The Great Turning. It is an in-between time. We may not make it. The dangers—both real and daunting—include nuclear annihilation, chemical poisoning, catastrophic climate change, starvation, epidemic disease outbreaks, and more. This cacophony of threats and wounds has been precipitated by an upstart species, *Homo sapiens*—us!

It is tempting to conclude that as a species, we have made a colossal mistake, fixated as we now are on constant growth and consumption. But maybe, just maybe, it will turn out not to have been a wrong turn after

THE AUDACITY TO IMAGINE

By the end of your lifetime you will be living in a world unimaginably more beautiful than the one you were born into. And it will be a world that is palpably improving year after year. . . . Prisons will no longer exist, and violence will be a rarity. Work will be about, "How may I best give of my gifts?" instead of "How can I make a living?" Crossing a national border will be an experience of being welcomed, not examined. Mines and quarries will barely exist, as we reuse the vast accumulation of materials from the industrial age. We will live in dwellings that are extensions of ourselves, eat food grown by people who know us and use articles that are the best that people in the full flow of their talents [can] make. . . . We will live in a richness of intimacy and community that hardly exists today. . . . And most of the time the loudest noises we hear will be the sounds of nature and the laughter of children.

Fantastical? The mind is afraid to hope for anything too good. If this description evokes anger, despair, or grief, then it has touched our common wound, the wound of separation. Yet the knowledge of what is possible lives on inside each of us, inextinguishable. Let us trust this knowing, hold each other in it and organize our lives around it. Do we really have any choice, as the old world falls apart? Shall we settle for anything less than a sacred world?

—Charles Eisenstein[3]

all; maybe the wacky frenzy of growth of the last two centuries, with all its inflicted wounds and wonders, has been necessary to get us to where we are going. Maybe it's all part of a developmental stage that our species has to go through?

In many respects, we are still a species in its adolescence. Like teenagers, in pursuit of action, power, and novelty, we often behave in ways that are self-indulgent and reckless.

Historically, the transformation from adolescence to adulthood was facilitated through a process of initiation wherein the young person, with the help of elders, was subjected to a series of trials and tribulations designed to catalyze profound awakenings. In effect, the adolescent's worldview fell apart and his identity was shattered. This temporary dissolution of ego created space for the young person to discover his/her place within society.

The amalgam of planetary crises that we now face can be seen as a call to *turn away from* our old, adolescence-imbued story of the separate self that exists to exploit and control the world and to *turn toward* a new story of the generative adult who sees the world with the eyes of kinship, compassion, and generosity. In this new story, our identity expands as we experience ourselves as *a part of* Earth's family of life.

Yes, at the deepest level, it seems that what we all yearn for is the lived experience of belonging to—being a part of—something greater than ourselves. The irony, of course, is that we already do belong to something that is colossally grand: We belong to Earth!

The visceral, embodied experience of being a part of Earth's body is the antidote to the *separation mind-set* that so often hijacks our consciousness.

We are all on a journey to become fully human. It is the journey of a lifetime. It is what gives our lives their meaning and purpose. It's why we are here.

Mis estimados: Do not lose heart. We were made for these times.[4]

Notes

Foreword

[1] David Orr, "Environmental Literacy: Education as if the Earth Mattered," in *Twelfth Annual E. F. Schumacher Lectures, People, Land, and Community*, 1993, page 1. http://www.sfsf.com.au/Education.As.If.The.Earth.Mattered.pdf

Preface to the Third Edition

[1] Vaclav Havel, quoted in Sharif Abdullah, *Creating a World That Works for All* (San Francisco: Berrett-Koehler, 1999), viii.

[2] This is not to ignore the fact that many noteworthy sustainability-related initiatives have been instituted over the past quarter century, especially in the realms of wind and solar power. See *Vital Signs: The Trends That Are Shaping Our Future*, vol. 22 (Washington, DC: The Worldwatch Institute; Island Press, 2015).

[3] Inspired, in part, by Bill Plotkin, *Wild Mind: A Field Guide to the Human Psyche* (Novato, CA: New World Library, 2013).

Part I: Earth, Our Home

[1] "BrainyQuote," https://www.brainyquote.com/quotes/marcel_proust_107111.

[2] Matthew Fox, *Coming of the Cosmic Christ* (San Francisco: Harper San Francisco, 1988), 32.

[3] Steve VanMatre, *Earth Education* (Greenville, WV: Institute for Earth Education, 1990).

Chapter 1: Humility

[1] Thomas Berry, *The Great Work* (New York: Bell Tower, 1999), 14–15.

[2] Chet Raymo, *Skeptics and True Believers* (New York: Walker, 1998), 243.

[3] Paul Krafel, *Seeing Nature: Deliberate Encounters with the Visible World* (White River Junction, VT: Chelsea Green, 1999).

[4] Krafel, *Seeing Nature*, 16.

[5] Brian Swimme, *The Hidden Heart of the Cosmos* (Maryknoll, NY: Orbis Books, 1996).

[6] Inspired by Johns Hopkins geologist George Fisher.

[7] Michael Seeds, *Horizons: Exploring the Universe* (Pacific Grove, CA: Brooks/Cole, 2002); and Chet Raymo, *An Intimate Look at the Night Sky* (New York: Walker, 2001).

[8] Raymo, *An Intimate Look at the Night Sky*.

[9] Phillip Ball, *A Biography of Water: Life's Matrix* (Berkeley: University of California Press, 2001); and Seeds, *Horizons*.

[10] Sarah Haleblian, "NASA Reveals First Picture of a Black Hole," *HS Insider/ Los Angeles Times*, June 12, 2019, https://highschool.latimes.com/la-canada -high-school/nasa-reveals-first-picture-of-a-black-hole/.

[11] Duane Elgin, *The Living Universe* (San Francisco: Berrett-Koehler Publishers, 2009).

[12] Brian Swimme and Mary Evelyn Tucker, *Journey of the Universe* (New Haven, CT: Yale University Press, 2011), 36.

[13] Chet Raymo, *Natural Prayers* (St. Paul, MN: Hungry Mind Press, 1999).

[14] Richard Heinberg, *The End of Growth* (Gabriola Island, BC: New Society Publishers, 2011).

[15] Lynn Margulis and Dorion Sagan, *Microcosmos* (Berkeley: University of California Press, 1997).

[16] David Suzuki, *The Sacred Balance* (Amherst, NY: Prometheus Books, 1998).

[17] Suzuki, *The Sacred Balance*; and Seeds, *Horizons*.

[18] Lynn Margulis, *Symbiotic Planet* (New York: Basic Books, 1998), 71–72.

[19] None of this rules out the possibility that there might be extraterrestrial beings in our galaxy that are unimaginably more complex and advanced than we are—beings that perhaps have already checked us out and concluded that we are not particularly interesting.

[20] Deepak Chopra, *The Book of Secrets* (New York: Harmony Books, 2004), 186.

[21] Swimme and Tucker, *Journey of the Universe*, 8–9.

[22] Brian Swimme, *The Universe Is a Green Dragon* (Santa Fe, NM: Bear, 1985).

[23] Christopher Uhl and Dana Stuchul, *Teaching as if Life Matters: The Promise of a New Education Culture* (Baltimore, MD: Johns Hopkins University Press, 2011).

[24] With the words *imperceptibly small* I am referring to the invisible, subatomic world that undergirds all of existence.

[25] Diarmuid O'Murchu, *Our World in Transition* (New York: Crossroad, 1992).

[26] Thomas Berry, *The Dream of the Earth* (San Francisco: Sierra Club Books, 1988), 123–24.

[27] Duane Elgin, *The Living Universe* (San Francisco: Berrett-Koehler Publishers, 2009), 8.

[28] Elgin, *The Living Universe*, 39.

[29] Elgin, *The Living Universe*, 39.

[30] Swimme and Tucker, *Journey of the Universe*, 2.

[31] This figure was inspired by a sketch in Seeds, *Horizons*, 231.

[32] Swimme, *The Hidden Heart of the Cosmos*, 52–53.

[33] Raymo, *An Intimate Look*, 6, 8.

Chapter 2: Curiosity and Connection

[1] Thomas Berry, *The Great Work, Our Way into the Future* (New York, Bell Tower, 1999), 4.

[2] Much of the following information on monarch butterfly natural history comes from Eric Grace, *The World of the Monarch Butterfly* (San Francisco: Sierra Club Books, 1997).

[3] Judith Kohl and Herbert Kohl, *The View from the Oak* (New York: New York Press, 1997).

4 Daniel Chamovitz, *What a Plant Knows: A Field Guide to the Senses* (New York: Scientific American/Farrar, Straus and Giroux, 2012).

5 U.S. Fish and Wildlife Service, "Save the Monarch: Overwintering Monarchs," https://www.fws.gov/midwest/monarch/overwinteringmonarchs.html.

6 The information on goldenrod ecology comes mainly from Arthur Weis and Warren Abrahamson, "Just Lookin' for a Home," *Natural History* 107 (1998): 60–63.

7 See Wikipedia, s.v. "*Demodex*," last modified August 23, 2019, http://en.wiki pedia.org/wiki/Demodex.

8 Lynn Margulis and Dorion Sagan, *What Is Life?* (Berkeley: University of California Press, 1995), 91.

9 Margulis and Sagan, *What Is Life?*, 90.

10 Jack Gilbert and Rob Knight, *Dirt Is Good* (New York: St. Martin's Press, 2017), 2.

11 Emeran Mayer, *The Mind-Gut Connection* (New York: HarperCollins, 2016).

12 Michael Morowitz, Erica Carlisle, and John Alverdy, "Contributions of Intestinal Bacteria to Nutrition and Metabolism in the Critically Ill," *Surgical Clinics of North America* 91, no. 4 (2011): 771–85.

13 Bruno Bonaz, Thomas Bazin, and Sonia Pellissier, "The Vagus Nerve at the Interface of the Microbiota-Gut-Brain Axis," *Frontiers in Neuroscience* 12 (2018): 49.

14 "How Bacteria Rule Over Your Body—The Microbiome," published October 5, 2017, YouTube video, https://www.youtube.com/watch?v=VzPD009qTN4.

15 Simon Carding, "Gut Bacteria and Mind Control," published May 5, 2015, YouTube video, https://www.youtube.com/watch?v=mioR_WrkRaU.

16 Shannon Harvey, "The Second Brain in Your Gut," The Connection, February 13, 2015, https://theconnection.tv/second-brain-gut.

17 Mayer, *The Mind-Gut Connection*.

18 Claudia Wallis, "How Gut Bacteria Help Make Us Fat and Thin," *Scientific American*, June 1, 2014, https://www.scientificamerican.com/article/how -gut-bacteria-help-make-us-fat-and-thin/.

19 Mayer, *The Mind-Gut Connection*, 18.

20 Knvul Sheikh, "How Gut Bacteria Tell Their Hosts What to Eat," *Scientific American*, April 25, 2017, https://www.scientificamerican.com/article/how -gut-bacteria-tell-their-hosts-what-to-eat/ and "How Bacteria Rule Over Your Body," YouTube video.

21 Wallis, "How Gut Bacteria Help Make Us Fat and Thin."

22 Ruairi Robertson, "Food for Thought," published December 7, 2015, YouTube video, https://www.youtube.com/watch?v=awtmTJW9ic8.

23 Albert Einstein, "Is the Universe Friendly?" http://www.awakin.org/read/view. php?tid=797.

24 Lynne McTaggart, *The Intention Experiment* (New York: Free Press, 2007).

25 Rupert Sheldrake, *The Sense of Being Stared At* (New York: Crown Publishers, 2003), 4.

26 Jeremy Hayward, *Letters to Vanessa* (Boston: Shambhala, 1997), 143–44.

27 Larry Dossey, *Healing Beyond the Body: Medicine and the Infinite Reach of the Mind* (Boston: Shambhala, 2003) (emphasis added).

28 Derrick Jensen, *A Language Older Than Words* (New York: Context Books, 2000), 126–27.

[29] The preceding three paragraphs were paraphrased from Christopher Uhl and Dana Stuchul, *Teaching as if Life Matters: The Promise of a New Education Culture* (Baltimore, MD: Johns Hopkins University Press, 2011), 165, and were also inspired by Anthony Weston, *Back to Earth* (Philadelphia: Temple University Press, 1994).

[30] Mr. Purrington, "Carl Jung: Often the Hands Know How to Solve a Riddle with Which the Intellect Has Wrestled in Vain," *Carol Jung Depth Psychology* (blog), February 2, 2019, https://carljungdepthpsychologysite.blog/2019/02/02/carl-jung-often-the-hands-know-how-to-solve-a-riddle-with-which-the-intellect-has-wrestled-in-vain-2/#.XTC7rVB7lBw.

[31] Inspired in part by Lucia Capacchione, *The Art of Emotional Healing* (Boston: Shambhala, 2006).

Chapter 3: Intimacy

[1] Thomas Berry, *The Dream of the Earth* (San Francisco: Sierra Club Books, 1988), 171.

[2] Joanna Macy and Molly Young Brown, *Coming Back to Life: Practices to Reconnect Our Lives, Our World* (Gabriola Island, BC: New Society Publishers, 1998).

[3] David Abram, *Becoming Animal: An Earthly Cosmology* (New York: Pantheon Books, 2010), 73, 77, 78.

[4] The Lewis Thomas quote is from Jeremy Hayward, *Letters to Vanessa* (Boston: Shambhala, 1997), 102–3; see the gut microbiome section in chapter 2 for more on the "wonders of the human body."

[5] Brian Swimme, *The Universe Is a Green Dragon* (Santa Fe, NM: Bear, 1984), 37.

[6] Tyler Volk, *Gaia's Body* (New York: Copernicus/Springer-Verlag, 1998).

[7] Paul Krafel, *Seeing Nature: Deliberate Encounters with the Visible World* (White River Junction, VT: Chelsea Green, 1999).

[8] Volk, *Gaia's Body.*

[9] Krafel, *Seeing Nature.*

[10] Evan Eisenberg, *The Ecology of Eden* (New York: Alfred A. Knopf, 1998).

[11] Eisenberg, *The Ecology of Eden*, 23.

[12] David Suzuki, *The Sacred Balance* (Amherst, NY: Prometheus Books, 1998), 8–9.

[13] Frank H. Bormann and Gene E. Likens, "Nutrient Cycling," *Science* 155 (1967): 424–29.

[14] It was assumed that the tiny bit of calcium lost each year was replenished by the natural weathering of rocks in the soil.

[15] Wolves sometimes track moose for long periods, eventually pushing them to exhaustion and then inflicting a fatal wound.

[16] David Abram, *Becoming Animal: An Earthly Cosmology* (New York: Pantheon Books, 2010), 59.

[17] Abram, *Becoming Animal.*

[18] The mosquito scenario is inspired by Abram, *Becoming Animal.*

[19] Abram, *Becoming Animal*, 62.

[20] Volk, *Gaia's Body.*

[21] Bernd Heinrich, *The Trees in My Forest* (New York: HarperCollins, 1997).

[22] Thich Nhat Hanh, *Peace Is Every Step* (New York: Bantam Books, 1991), 95.

23 J. Lee, "Honoring the Given World: An Interview with Scott Russell Sanders," in *Stonecrop* (Denver: River Lee Book Company, 1997), 29.

24 Daily's thought experiment was inspired by John Holdren and is described in Andrew Beattie and Paul Ehrlich, *Wild Solutions* (New Haven, CT: Yale University Press, 2001).

25 Beattie and Ehrlich, *Wild Solutions*.

26 Brian Swimme, *The Hidden Heart of the Cosmos* (Maryknoll, NY: Orbis Books, 1996), 56.

27 Richard Louv, *Last Child Left in the Woods* (New York: Workman Publishing Company, 2005). For more recent data see Jane Wakefield, "Children Spend Six Hours or More a Day on Screens," BBC News, March 27, 2015, https://www.bbc.com/news/technology-32067158.

28 Chellis Glendinning, "Technology, Trauma, and the Wild," in *Ecopsychology: Restoring the Earth, Healing the Mind*, ed. Theodore Rozak, Mary Gomes, and Allen Kanner (San Francisco: Sierra Club Books, 1995), 41.

29 Cited in Louv, *Last Child Left in the Woods*.

30 Louv, *Last Child Left in the Woods*.

31 Stan Rowe, "Education for a New World View," Trumpeter, 1991, http://trumpeter.athabascau.ca/index.php/trumpet/article/view/455/752.

32 I was first introduced to this practice through Project NatureConnect (www.ecopsych.com).

33 Jon Kabat-Zinn, *Full Catastrophe Living* (New York: Dell, 1990), 27–28 (very slightly paraphrased).

34 Steve Van Matre, *Earth Education: A New Beginning* (Greenville, WV: Institute for Earth Education, 1990).

Part II: Assessing the Health of Earth

1 Miguel Barrientos, "How Many People Die a Day in the U.S.?," *IndexMundi Blog*, March 5, 2018, https://www.indexmundi.com/blog/index.php/2018/03/05/how-many-people-die-a-day-in-the-us/.

2 "World Scientists' Warning to Humanity" (Boston: Union of Concerned Scientists, 1992), www.ucsusa.org.

Chapter 4: Listening

1 Results from Google search, "When you talk, you are only repeating what you already know. . . . ," https://www.google.com/search?q=When+you+talk,+you+are+only+repeating+what+you+already+know.+But+if+you+listen,+you+may+learn+something+new.&client=firefox-b-1&tbm=isch&source=iu&ictx=1&fir=rP9gOeFjQzRvkM%253A%252Cv_0mJSmSCe64pM%252C_&usg=AI4_-kRiZn5_JD71d4y9RZW9RL9vc10tlg&sa=X&ved=2ahUKEwii37ndwvffAhUNMawKHRrmeD60Q9QEwAnoECAUQCA#imgrc=_.

2 For more information see https://cumulis.epa.gov/supercpad/SiteProfiles/index.cfm?fuseaction=second.cleanup&id=0300444.

3 Though I focus on bird migration in the Americas, it is also true that many birds that breed in Europe migrate to Africa in the winter.

4 John Terborgh, "Why American Songbirds Are Vanishing," *Scientific American*, May 1992, 98–104.

5 Presently this survey is administered by the U.S. Geological Survey in conjunction with the Canadian Wildlife Survey.

6 Terborgh, "Why American Songbirds Are Vanishing." See also J. R. Sauer, J. E. Hines, J. E. Fallon, K. L. Pardieck, D. J. Ziolkowski Jr., and W. A. Link, *The North American Breeding Bird Survey, Results and Analysis 1966–2009*, Version 3.23.2011 (Laurel, MD: USGS Patuxent Wildlife Research Center, 2011), www.mbr-pwrc.usgs.gov/bbs/bbs2009.html; Lauren McDonald, "Report Shows Steep Declines for North American Birds," The Wildlife Society, May 24, 2016, http://wildlife.org/report-shows-steep-declines-for-north-american-birds/; and "State of North America's Birds 2016," Cornell University, http://www.stateof thebirds.org/2016/overview/results-summary/.

7 Chuck Raasch, "Cats Kill up to 3.7B Birds Annually," *USA Today*, January 29, 2013, https://www.usatoday.com/story/news/nation/2013/01/29/cats-wild -birds-mammals-study/1873871/.

8 David Wilcove, "Nest Predation in Forest Tracts and the Decline of Migratory Songbirds," *Ecology* 66 (1985): 1211–14; see also Terborgh, "Why American Songbirds Are Vanishing."

9 Christian Both, Sandra Bouwhuis, C. M. Lessells, and Marcel Visser, "Climate Change and Population Declines in a Long-Distance Migratory Bird," *Nature*, May 2006, 81–83; see also Florida Museum of Natural History, "Migratory Birds Bumped Off Schedule as Climate Change Shifts Spring," *Science Daily*, May 15, 2017, https://www.sciencedaily.com/releases/2017/05/170515091126.htm.

10 Stuart Pimm, *The World According to Pimm: A Scientist Audits the Earth* (New York: McGraw-Hill, 2001).

11 "Summary Statistics," The IUCN Red List of Threatened Species, 2018, https:// www.iucnredlist.org/.

12 Pimm, *The World According to Pimm.*

13 I have borrowed this analogy from Paul Ehrlich and Ann Ehrlich, *Extinction* (New York: Random House, 1981).

14 The failed Biosphere 2 experiment, described in the previous chapter, is a reminder that Earth's ecosystems are far more complex than we are presently able to understand.

15 Aldo Leopold, *A Sand County Almanac* (Oxford: Oxford University Press, 1949).

16 Sylvia Earle, *The World Is Blue* (Washington, DC: National Geographic Society, 2009).

17 Earle, *The World Is Blue*, 15.

18 Tobin Hart, *The Secret Spiritual World of Children* (Novato, CA: New World Library, 2003), 47.

19 Deborah Cramer, *Smithsonian Ocean: Our Water Our World* (Washington, DC: Smithsonian Institution with HarperCollins, 2008), 114.

20 Earle, *The World Is Blue*, 17.

21 Darlene Crist, Gail Scowcroft, and James M. Harding, *World Ocean Census* (Buffalo, NY: Firefly Books, 2009).

22 Cramer, *Smithsonian Ocean.*

[23] This idea of a journey from ocean surface to ocean depths was inspired by Earle, *The World Is Blue*.

[24] Cramer, *Smithsonian Ocean*.

[25] Results from Google search, "statistics on annual ocean fish catch," https://www .google.com/search?client=firefox-b-1&q=statistics+on+annual+ocean+fish+catch.

[26] Wikipedia, s.v. "World fisheries production," last modified April 24, 2019, https:// en.wikipedia.org/wiki/World_fisheries_production; see also Food and Agriculture Organization of the United Nations, *The State of World Fisheries and Aquaculture 2016* (Rome: FAO, 2016), http://www.fao.org/3/a-i5555e.pdf.

[27] Though these species are not threatened with extinction at the present time, if current unsustainable fishing practices continue, their populations will decline.

[28] Peter Brannen, *The Ends of the World: Volcanic Apocalypses, Lethal Oceans, and Our Quest to Understand Earth's Past Mass Extinctions* (New York: Ecco Books, 2017), 235.

[29] Data assembled by the World Wildlife Fund reveal that upward of 300,000 marine mammals and hundreds of thousands of turtles and seabirds are inadvertently killed each year in industrial fishing operations.

[30] Tierney Thys, "For the Love of Fishes," in *Oceans: The Threats to Our Seas and What You Can Do to Turn the Tide*, ed. Jon Bowermaster (Philadelphia: Perseus Book Group, 2010), 140.

[31] Wilma Subra, "Reasons to Worry about the Dead Zones," in Bowermaster, *Oceans*, 131.

[32] "Average-sized Dead Zone Forecasted for the Gulf of Mexico," USGS, June 7, 2018, https://www.usgs.gov/news/average-sized-dead-zone-forecasted-gulf -mexico.

[33] Damian Carrington, "Oceans Suffocating as Huge Dead Zones Quadruple since 1950, Scientists Warn," *Guardian*, January 4, 2018, https://www.theguardian .com/environment/2018/jan/04/oceans-suffocating-dead-zones-oxygen -starved; and Earle, *The World Is Blue*.

[34] Carrington, "Oceans Suffocating"; see also L. Lebreton et al., "Evidence That the Great Pacific Garbage Patch Is Rapidly Accumulating Plastic," *Scientific Reports* 8 (March 22, 2018), https://www.nature.com/articles/s41598-018-22939-w.

[35] Results from Google search, "plastic reaching the ocean each year," https://www. google.com/search?client=firefox-b-1&ei=_BVCXMDHLIicjwStjKX4CQ&q =plastic+reaching+the+ocean+each+year&oq=plastic+reaching+the+ocean +each+year&gs_l=psy-ab.3..33i22i29i30.1584292.1593738..1594076 ...4.0..0.94.2970.40......0....1..gws-wiz.......0i71j0i67j0j0i131j35i39j0i131i67j0 i20i263j0i10j0i22i30.eBurKic8SLc.

[36] Earle, *The World Is Blue*.

[37] David de Rothschild, "Message in a Bottle: Adventures Aboard the Plastiki," in Bowermaster, *Oceans*, 83.

[38] Susan Casey, "Plastic Ocean: Our Oceans Are Turning into Plastic . . . Are We?," in Bowermaster, *Oceans*, 71; see also results from Google search, "americans per capita plastic waste," https://www.google.com/search?client=firefox-b-1&q=ameri cans+per+capita+plastic+waste.

[39] "How to Improve the Health of the Oceans," *Economist*, May 27, 2017.

40 Earle, *The World Is Blue*, 112.

41 Alejandra Borunda, "This Young Whale Died with 88 Pounds of Plastic in Its Stomach," *National Geographic*, March 18, 2019, https://www.nationalgeo graphic.com/environment/2019/03/whale-dies-88-pounds-plastic-philip pines/; see also "Plastic in Our Oceans Is Killing Marine Mammals," World Wildlife Fund, October 11, 2018, https://www.wwf.org.au/news/blogs/plas tic-in-our-oceans-is-killing-marine-mammals#gs.KGtFMVl5.

42 Chelsea M. Rochman, Eunha Hoh, Brian T. Hentschel, and Shawn Kaye. "Long-Term Field Measurement of Sorption of Organic Contaminants to Five Types of Plastic Pellets: Implications for Plastic Marine Debris," *Environmental Science & Technology* 47, no. 3 (2013): 1646–54.

43 Mike McCrae, "99% of Ocean Plastic Waste Is Invisible, But This Method Could Help Find It," Science Alert, November 24, 2017, https://www.science alert.com/missing-99-percent-ocean-microplastic-detection-method.

44 Ingrid Bacci, *The Art of Effortless Living: Simple Techniques for Healing Mind, Body and Spirit* (Croton-on-Hudson, NY: Vision Works, 2000), 110.

45 Joanna Macy and Molly Brown, *Coming Back to Life: Practices to Reconnect Our Lives, Our World* (Gabriola Island, BC: New Society Publishers, 1998).

46 Macy and Brown, *Coming Back to Life*, 27.

47 Paraphrased from Christopher Uhl and Dana Stuchul, *Teaching as if Life Matters: The Promise of a New Education Culture* (Baltimore, MD: Johns Hopkins University Press, 2011).

48 Joanna Macy and Molly Brown give details on this "Truth Mandala" ritual in their book *Coming Back to Life*.

49 Bacci, *The Art of Effortless Living*, 116.

50 Diarmuid O'Murchu, *Our World in Transition* (New York: Crossroad, 1992), 152.

51 For a fuller elaboration, see the discussion of speciesism in chapter 2.

52 This figure was inspired by Michael McKinney and Robert Schoch, *Environmental Science: Systems and Solutions* (New York: West, 1996), 596–97.

53 "Karen A. Anderson, Quotes, Quotable Quote," https://www.goodreads.com/quotes/8924972-learning-how-to-communicate-with-animals-is-just-like-learning.

54 Barbara Smuts, "Encounters with Wild Minds," *Journal of Consciousness Studies* 8, nos. 5–7 (2001): 293–309.

55 See Non-Judgment Day Project, www.nonjudgmentday.org/index.html.

Chapter 5: Courage

1 Wes Nisker, *The Essential Crazy Wisdom* (Berkeley, CA: Ten Speed Press, 2001), 13.

2 Scripps CO_2 Program, "Mauna Loa Record," http://scrippsco2.ucsd.edu/graph ics_gallery/mauna_loa_record/mauna_loa_record.html.

3 Tyler Volk, *Gaia's Body* (New York: Copernicus/Springer-Verlag, 1998).

4 Bill McKibben, *Eaarth: Making a Life on a Tough New Planet* (New York: Henry Holt, 2010), 27–28.

5 McKibben, *Eaarth*, 27–28.

6 Bill McKibben, *Falter: Has the Human Game Begun to Play Itself Out?* (New York: Henry Holt, 2019), 13.

[7] Bill McKibben, *The End of Nature* (New York: Anchor Books, 1989), 18.

[8] Just as percent means *out of a hundred*, parts per million (ppm) means *out of a million*.

[9] Climate Central, "Hottest Years on Record Globally," https://www.climatecentral.org/gallery/graphics/the-10-hottest-global-years-on-record.

[10] Ciara Nugent, "Carbon Dioxide Concentration in the Earth's Atmosphere Has Hit Levels Unseen for 3 Million Years," *Time*, May 14, 2019, https://time.com/5588794/carbon-dioxide-earth-climate-change/.

[11] Eric Holthaus, "It's (Nearly) Official: 2018 Was the 4th Warmest Year in Recorded History," *Grist*, December 21, 2018, https://grist.org/article/its-nearly-official-2018-was-the-4th-warmest-year-in-recorded-history/.

[12] The IPCC was established by the UN in 1988 and is charged with integrating research findings from more than two thousand climate scientists representing more than 120 nations. Intergovernmental Panel on Climate Change, *Summary for Policymakers of IPCC Special Report on Global Warming of 1.5°C Approved by Governments*, October 8, 2018, https://www.ipcc.ch/2018/10/08/summary-for-policymakers-of-ipcc-special-report-on-global-warming-of-1-5c-approved-by-governments/.

[13] Greg Lankenau provided this stereo analogy, as well as the beach phrasing.

[14] Gayathri Vaidyanathan, "Killer Heat Grows Hotter around the World," *Scientific American*, August 6, 2015.

[15] Bob Berwyn, "Heat Waves Creeping Toward a Deadly Heat-Humidity Threshold," InsideClimate News, August 3, 2017.

[16] Wikipedia, s.v. "Climate change in the Arctic," last modified October 8, 2019, https://en.m.wikipedia.org/wiki/Climate_change_in_the_Arctic

[17] Jeremy White and Kendra Pierre-Louis, "In the Arctic, the Old Ice Is Disappearing," *New York Times*, May 14, 2018, https://www.nytimes.com/interactive/2018/05/14/climate/arctic-sea-ice.html.

[18] Sylvia Earle, *The World Is Blue: How Our Fate and the Ocean's Are One* (Washington, DC: National Geographic Society, 2009).

[19] It is also worth noting that a 2012 study reported that the enormous ice sheets of Greenland and Antarctica had increased their rate of melting threefold between 1990 and 2010. Jet Propulsion Laboratory, "Ice Sheet Loss at Both Poles Increasing, Study Finds," November 29, 2012, www.jpl.nasa.gov/news/news.php?release=2012-376&cid=release_2012-376.

[20] Bridget Alex, "Artic Meltdown: We're Already Feeling the Consequences of Thawing Permafrost," *Discover*, June 2018, http://discovermagazine.com/2018/jun/something-stirs.

[21] McKibben, *Falter*.

[22] Maddie Stone, "Arctic Wildfires Are Releasing as Much Carbon as Belgium Did Last Year," *Grist*, August 2, 2019, https://grist.org/article/the-arctic-is-having-unprecedented-wildfires-heres-why-that-matters/?utm_medium=email&utm_source=newsletter&utm_campaign=daily. See also Michelle C. Mack, M. Syndonia Bret-Harte, Teresa N. Hollingsworth, Randi R. Jandt, Edward A. G. Schuur, Gaius R. Shaver, and David L. Verbyla, "Carbon Loss from an Unprecedented Arctic Tundra Wildfire," *Nature* 475 (2011): 489–92; and Tom Di Liberto, "Wildfire Still Burning in Greenland Tundra in Mid-August 2017," Climate.gov,

August 18, 2017, https://www.climate.gov/news-features/event-tracker/wild fire-still-burning-greenland-tundra-mid-august-2017.

23 A. York, U. Bhatt, R. Thoman, and R. Ziel, "Wildland Fire in High Latitudes," *Arctic Report Card: Update for 2017*, December 5, 2017, https://arctic.noaa .gov/Report-Card/Report-Card-2017/ArtMID/7798/ArticleID/692/Wild land-Fire-in-High-Latitudes.

24 Joe Romm, "The Permafrost Won't Be *Perma* for Long," *Think Progress*, May 22, 2008, http://thinkprogress.org/romm/2008/05/22/202419/tundra -part-1-the-permafrost-wont-be-perma-for-long/; and Joe Romm, "For Peat's Sake: A Point of No Return as Alarming as the Tundra Feedback," Climate Progress.org, October 1, 2008.

25 Rebecca Lindsey, "Climate Change: Global Sea Level," Climate.gov, September 19, 2019, https://www.climate.gov/news-features/understanding-climate/cli mate-change-global-sea-level.

26 Eric Holthaus, "Don't Mean to Alarm You, but There's a Big Hole in the World's Most Important Glacier," *Grist*, January 31, 2019, https://grist.org/ article/dont-mean-to-alarm-you-but-theres-a-big-hole-in-the-worlds-most -important-glacier/?utm_medium=email&utm_source=newsletter&utm_cam paign=daily.

27 Eric Holthaus, "Doomsday Postponed? What to Take from the Big New Antarctica Studies," *Grist*, February 7, 2019, https://grist.org/article/doomsday-postponed -what-to-take-from-the-big-new-antarctica-studies/?utm_medium=email&utm_ source=newsletter&utm_campaign=daily.

28 Pavel Y. Groisman, Richard W. Knight, and Thomas R. Karl, "Heavy Precipitation and High Streamflow in the Contiguous United States: Trends in the Twentieth Century," *Bulletin of the American Meteorological Society* 82, no. 2 (2001): 219.

29 Tom Knudson, "Sierra Nevada Climate Changes Feed Monster, Forest-Devouring Fires," *Sacramento Bee*, November 30, 2008.

30 Jim Robbins, "Bark Beetles Kill Millions of Acres of Trees in the West," *New York Times*, November 18, 2008.

31 "Wildfire Statistics," Congressional Research Service, September 3, 2019, https:// fas.org/sgp/crs/misc/IF10244.pdf.

32 Michael Kodas, *Megafire: The Race to Extinguish a Deadly Epidemic of Flame* (Boston: Houghton Mifflin Harcourt, 2017), xii.

33 Kodas, *Megafire*, 20.

34 "Deathwatch for the Amazon," *Economist*, August 1, 2019, https://www.econ omist.com/leaders/2019/08/01/deathwatch-for-the-amazon.

35 Sarah Gibbens, "The Amazon Is Burning at Record Rates—and Deforestation Is to Blame," *National Geographic*, August 21, 2019, https://www.national geographic.com/environment/2019/08/wildfires-in-amazon-caused-by -deforestation/.

36 "Rate of Ocean Warming Quadrupled Since Late 20th Century, Study Reveals," *Carbon Brief*, March 10, 2017.

37 Deborah Cramer, *Smithsonian Ocean: Our Water, Our World* (New York: Smithsonian in association with HarperCollins, 2008).

[38] Lisa Suatoni, "Acid Test," in *Oceans: The Threats to Our Seas and What You Can Do to Turn the Tide*, ed. Jon Bowermaster (Philadelphia: Perseus Book Group, 2010), 103–6.

[39] Eelco Rohling, *The Oceans: A Deep History* (Princeton, NJ: Princeton University Press, 2017), 181.

[40] Earle, *The World Is Blue*, 179.

[41] McKibben, *Eaarth*, xiii.

[42] McKibben, *Falter*, 72.

[43] McKibben, *Falter*, 72.

[44] McKibben, *Falter*, 74.

[45] Benjamin Franta, "Shell and Exxon's Secret 1980s Climate Change Warnings, *Guardian*, September 19, 2018.

[46] McKibben, *Falter*, 72.

[47] Jason M. Breslow, "Investigation Finds Exxon Ignored Its Own Early Climate Change Warnings," PBS.org, September 16, 2015.

[48] McKibben, *Falter*, 76.

[49] Ruairi Arrieta-Kenna, "Almost 90% of Americans Don't Know There's Scientific Consensus on Global Warming," vox.com, July 6, 2017.

[50] Alex Steffen, "On Climate, Speed Is Everything," *The Nearly Now*, December 7, 2017.

[51] I wish to acknowledge Joann Macy for this framing.

[52] Soil erosion data are for the U.S. cropland average for nonfederal rural land, accessed at www.nrcs.usda.gov.

[53] David Pimentel, "Soil Erosion: A Food and Environmental Threat," *Journal of the Environment, Development and Sustainability* 8 (2006): 119–37.

[54] Pimentel, "Soil Erosion."

[55] Water Encyclopedia, s.v. "Ogallala Aquifer," www.waterencyclopedia.com/Oc-Po/Ogallala-Aquifer.html.

[56] Theo Colburn, "Toxic Legacy," *Yes!*, Summer 1998, 14–18.

[57] "Chemicals in the Human Body," The World Counts, thttp://www.theworldcounts.com/counters/impact_of_pollution_on_human_health/chemicals_in_the_human_body.

[58] Sandra Steingraber, public lecture delivered at Penn State University, October 23, 2000.

[59] Steingraber, public lecture.

[60] Though Scorecard is still accessible on the web, it has not been updated for some time.

[61] Sandra Steingraber, "PVC and the Breasts of Mothers," *Rachel's Environment and Health*, July 1999, www.rachel.org/?q=en/node/4833.

[62] Sandra Steingraber, *Living Downstream: A Scientist's Personal Investigation of Cancer and the Environment* (New York: Vintage Books, 1998).

[63] This figure was inspired by Theo Colburn, Diane Dumanoski, and J. Peter Myers, *Our Stolen Future* (New York: Dutton, 1996), 72.

[64] Colburn et al., *Our Stolen Future*.

[65] "Study Confirms Estrogen in Water from the Pill Devastating to Fish Population," Lifesite News, February 18, 2008, www.lifesitenews.com/news/archive/ldn/2008/feb/08021805.

66 The Endocrine Disruption Exchange home page, https://endocrinedisruption .org/.

67 Laura Vandenberg, Theo Colburn, Tyrone Hayes, Jerrold Heindel, David Jacobs Jr., Duk-Hee Lee, Toshi Shioda, Ana Soto, Frederick vom Saal, Wade Welshons, Thomas Zoeller, and John Myers, "Hormones and Endocrine-Disrupting Chemicals: Low-Dose Effects and Nonmonotonic Dose Responses," *Endocrine Review* 33 (2012): 378–455.

68 Thaddeus Schug, Amanda Janesick, Bruce Blumberg, and Jerrid Heindel, "Endocrine Disrupting Chemicals and Disease Susceptibility," *Journal of Steroid Biochemistry and Molecular Biology* 127 (2011): 204–15.

69 Nicholas D. Kristof, "It's Time to Learn from Frogs," *New York Times*, June 27, 2009.

70 Hagai Levine, Niels Jørgensen, Anderson Martino-Andrade, Jaime Mendiola, Dan Weksler-Derri, Irina Mindlis, Rachel Pinotti, and Shanna H. Swan, "Temporal Trends in Sperm Count: A Systematic Review and Meta-regression Analysis," *Human Reproduction Update* 23, no. 6 (2017): 646–59.

71 Derrick Jensen, *A Language Older Than Words* (New York: Context Books, 2000), 2.

72 Clive Hamilton, *Defiant Earth: The Fate of Humans in the Anthropocene* (Cambridge, UK: Polity Press, 2017), 42.

73 Charles Eisenstein, *Climate A New Story* (Berkeley, CA: North Atlantic Books, 2019).

74 David Bohm, Donald Factor, and Peter Garrett, "Dialogue: A Proposal," www .david-bohm.net/dialogue/dialogue_proposal.html.

75 Tamarack Song, *Sacred Speech: The Way of Truthspeaking* (Three Lakes, WI: Teaching Drum Outdoor School, 2004), i.

76 Inspired by Austin Mandryk.

Chapter 6: Living the Questions

1 Neil Postman and Charles Weingartner, *Teaching as a Subversive Activity* (New York: Delta, 1969), 23.

2 For an overview of this research, see C. Uhl, "You Can Keep a Good Forest Down," *Natural History Magazine* 92 (1983): 70–79; and C. Uhl, "Restoration of Degraded Lands in the Amazon Basin," in *Biodiversity*, ed. E. O. Wilson (Washington, DC: National Academy Press, 1988).

3 Fritz Stern, "The Importance of 'Why,'" *World Policy Journal*, Spring 2000, 1–8.

4 United Nations Department of Economic and Social Affairs, "World Population Projected to Reach 9.8 Billion in 2050, and 11.2 Billion in 2100," June 21, 2017, https://www.un.org/development/desa/en/news/population/world-popu lation-prospects-2017.html.

5 Bill McKibben, *Deep Economy: The Wealth of Communities and the Durable Future* (New York: Henry Holt, 2007); "Kerala: Human Development Report," State Planning Board, Government of Kerala, 2006; and Wikipedia, s.v. "Kerala," last modified October 9, 2019, http://en.wikipedia.org/wiki/Kerala#Hu man_Development_Index.

6 Bill McKibben, "The Enigma of Kerala," *Utne Reader*, March–April 1996, 103–11.

7 McKibben, "The Enigma of Kerala."

8 Wikipedia, s.v. "Kerala"; see also "Population Development: What Kerala Can Teach India and China," Soapboxie, January 31, 2019, https://soapboxie .com/world-politics/Population-Development-What-Kerala-can-Teach -India-and-China.

9 Ellen J. Langer, *Mindfulness* (Cambridge, MA: Perseus Books, 1989), 9.

10 Sources for graphs: population, Matt Rosenberg, "Current World Population and Future Projections," ThoughtCo., July 19, 2019, http://geography.about .com/od/obtainpopulationdata/a/worldpopulation.htm; gross world product, Wikipedia, s.v. "Gross world product," last modified August 14, 2019, http:// en.wikipedia.org/wiki/Gross_world_product; water, "Water—The New Oil," Chess.com, www.chess.com/forum/view/off-topic/water—the-new-oil?page=4; paper, Sam Martin, "Paper Chase," Ecology, September 10, 2011, www.ecology .com/2011/09/10/paper-chase/; motor vehicles, Wikipedia, s.v. "Motor vehicle," last modified September 21, 2019, http://en.wikipedia.org/wiki/Motor_ vehicle; mobile phones, Wikipedia, s.v. "Mobile phone," last modified October 1, 2019, http://en.wikipedia.org/wiki/Mobile_phone.

11 Juliet Schor, *Plentitude* (New York: Penguin Press, 2010).

12 Climate News Network, "Human Consumption of Earth's Natural Resources Has Tripled in 40 Years," EcoWatch, July 25, 2016, https://www.ecowatch .com/humans-consumption-of-earths-natural-resources-tripled-in-40-years -1943126747.html.

13 Schor, *Plentitude*.

14 James Speth, *The Bridge at the Edge of the World: Capitalism, the Environment, and Crossing from Crisis to Sustainability* (New Haven, CT: Yale University Press, 2008); and Schor, *Plentitude*.

15 Results from Google search "how many cell phones do u.s. people discard/ year," https://www.google.com/search?client=firefox-b-1-d&q=how+many+cell +phones+do+u.s.+people+discard%2Fyear.

16 Richard Layard, *Happiness: Lessons from a New Science* (New York: Penguin Press, 2005).

17 A. Guttmann, "Advertising Spending in the United States from 2010 to 2019 (in Billion U.S. Dollars)," Statista, August 9, 2019, https://www.statista.com/statis tics/236958/advertising-spending-in-the-us/.

18 Pacific Institute, "Bottled Water and Energy: A Fact Sheet," www.pacinst.org/top ics/water_and_sustainability/bottled_water/bottled_water_and_energy.html.

19 Emily Arnold and Janet Larsen, "Bottled Water: Pouring Resources Down the Drain," Earth Policy Institute, February 2, 2006, www.earthpolicy.org/ index.php?/plan_b_updates/2006/update51; see also results from Google search "bottled water from tap?," https://www.google.com/search?client=fire fox-b-1-d&q=bottled+water+from+tap%3F.

20 National Resources Defense Council, *Bottled Water: Pure Drink or Pure Hype?*, www.nrdc.org/water/drinking/bw/exesum.asp.

21 Results from Google search "percentage of water bottles that are recycled," https://www.google.com/search?client=firefox-b-1-d&q=percentage+of+wa ter+bottles+that+are+recycled; see also David de Rothschild, "Message in a Bottle: Adventures Aboard the Plastiki," in *Oceans: The Threats to Our Seas and What You*

Can Do to Turn the Tide, ed. Jon Bowermaster (New York: PublicAffairs, 2010), 83–91.

22 Andrew Theen, "Ivy Colleges Shunning Bottled Water Jab at $22 Billion Industry," Bloomberg.com, March 7, 2012, www.bloomberg.com/news/2012-03 -07/ivy-colleges-shunning-bottled-water-jab-at-22-billion-industry.html.

23 Energy constitutes the lion's share of most people's footprint. Analysts calculate the energy footprint of a product by first determining the amount of fossil fuel needed to make the product along with the quantity of CO_2 released in the burning of that fossil fuel. Then they calculate the land area required to absorb the said amount of CO_2. This approach is taken because the unrestricted release of carbon dioxide into the atmosphere has begun to destabilize Earth's climate (see chapter 5). Hence, to stabilize global climate, land must be set aside to assimilate the carbon dioxide released in the burning of fossil fuels.

24 Chelsea Harvey, "We Are Killing the Environment One Hamburger at a Time," Business Insider, March 5, 2015, https://www.businessinsider.com/one-ham burger-environment-resources-2015-2.

25 Wikipedia, s.v. "List of Countries by Ecological Footprint," last modified August 29, 2019, http://en.wikipedia.org/wiki/List_of_countries_by_ecological_footprint.

26 Global Footprint Network, *Global Footprint Network 2007 Annual Report*, www. footprintnetwork.org.

27 Global Footprint Network, "Overshoot Trends," www.footprintnetwork.org/en/ index.php/GFN/page/overshoot_trends.

28 The Limits to Growth study was revised in 2004 with generally similar findings. Donella Meadows, Jorgen Randers, and Dennis Meadows, *Limits to Growth: The 30-Year Update* (White River Junction, VT: Chelsea Green Publishers, 2004).

29 Jared Diamond, *Collapse* (New York: Viking, 2005), 504.

30 Results from Google search "gus speth quotes . . . keep on doing," https:// www.google.com/search?q=gus+speth+quotes...+keep+on+doing&client=fire fox-b-1&tbm=isch&source=iu&ictx=1&fir=Q2ESND_2am0FzM%253A %252CPfqHE2njL6P9qM%252C_&usg=AI4_-kSVmLXFnXl2Yd4PfR4TZvJi sho0sg&sa=X&ved=2ahUKEwiEwKbQ65XgAhUnzoMKHVhKCSkQ9QEwA noECAUQBg#imgrc=VBe0pCyqMTL8KM.

31 Happy Planet Index, www.happyplanetindex.org/.

32 This framing was suggested to me by Gregory Lankenau.

33 Wendell Berry, *The Long-Legged House* (New York: Harcourt, Brace & World, 1969).

34 Martha Beck, *Finding Your Own North Star: Claiming the Life You Were Meant to Live* (New York: Three Rivers, 2001), 162.

35 "11 Facts about Global Poverty," DoSomething.org, https://www.dosome thing.org/us/facts/11-facts-about-global-poverty.

36 Rainer Maria Rilke, *Letters to a Young Poet*, trans. M. D. Herter Norton (New York: W. W. Norton, 1993), 33, 35.

37 Jerry Mander, *In the Absence of the Sacred: The Failure of Technology and the Survival of the Indian Nations* (San Francisco: Sierra Club Books, 1991), 43–44.

38 Derrick Jensen, *A Language Older Than Words* (New York: Context Books, 2000), 40.

Part III: Healing Ourselves, Healing Earth

[1] Jelaluddin Rumi, in *Open Secret*, trans. John Moyne and Coleman Barks (Boston: Shambhala), 7.

[2] Eckhart Tolle, *The Power of Now* (Novato, CA: New World Library, 1999), 67.

Chapter 7: The Old Story

[1] Quoted in Sabrina Hassumani, *Salman Rushdie: A Postmodern Reading of His Major Works* (Cranbury, NJ: Associated University Presses, 2002), 104.

[2] David Orr, *Earth in Mind* (Washington, DC: Island Press, 1994), 12.

[3] Charles Eisenstein, *The Ascent of Humanity* (Harrisburg, PA: Panenthea Press, 2007); Joseph Chilton Pearce, *The Biology of Transcendence: A Blueprint of the Human Spirit* (Rochester, VT: Park Street Press, 2002); and Bill Plotkin, *Nature and the Human Soul: Cultivating Wholeness and Community in a Fragmented World* (Novato, CA: New World Library, 2008).

[4] Daniel Quinn, *Ishmael* (New York: Bantam, Turner, 1992), 35–36.

[5] Barbara M. Hubbard, *Conscious Evolution: Awakening the Power of Our Social Potential* (Novato, CA: New World Library, 1998), 178.

[6] Quinn, *Ishmael*.

[7] Quinn, *Ishmael*.

[8] Christopher Uhl and Dana Stuchul, *Teaching as if Life Matters: The Promise of a New Education Culture* (Baltimore, MD: Johns Hopkins University Press, 2011).

[9] Barbara Brandt, *Whole Life Economics* (Gabriola Island, BC: New Society Publishers, 1995).

[10] Paraphrased from Susan Strauss, *The Passionate Fact* (Golden, CO: North American Press, 1996), 9.

[11] Brandt, *Whole Life Economics*.

[12] Jerry Mander, *In the Absence of the Sacred: The Failure of Technology and the Survival of the Indian Nations* (San Francisco: Sierra Club Books, 1991).

[13] Fareed Zakaria, "The Try-Hard Generation," *Atlantic*, June 10, 2015, https://www.theatlantic.com/education/archive/2015/06/in-defense-of-a-try-hard-generation/394535/.

[14] Harwood Group, "Yearning for Balance," *Yes!*, Spring/Summer 1996, 17.

[15] Herman Daily, *From Uneconomic Growth to a Steady-State Economy* (Northampton, MA, Edward Elgar Publishing, 2014); see also "What Is a Steady State Economy?," Center for the Advancement of the Steady State Economy, https://steadystate.org/wp-content/uploads/CASSE_Brief_SSE.pdf.

[16] Bill McKibben, "Recalculating the Climate Math," *New Republic*, September 22, 2016, https://newrepublic.com/article/136987/recalculating-climate-math.

[17] McKibben, "Recalculating the Climate Math."

[18] Charles Eisenstein, *Sacred Economics: Money, Gift and Society in the Age of Transition* (Berkeley, CA: Evolver Editions, 2011).

[19] Time spent on screens from "People Spend Most of Their Waking Hours Staring at Screens," MarketWatch, August 4, 2018, https://www.marketwatch.com/story/people-are-spending-most-of-their-waking-hours-staring-at-screens-2018-08-01; time spent in socializing and face-to-face conversation

from Bureau of Labor Statistics, "American Time Use Survey—2018 Results," news release, June 19, 2019, https://www.bls.gov/news.release/pdf/atus.pdf.

20 Juliet Schor, *Plentitude* (New York: Penguin Press, 2010).

21 Quoted in John de Graaf, ed., *Take Back Your Time* (San Francisco: Berrett-Kohler Publishers, 2003).

22 "Average Size of US Homes, Decade by Decade," Newser, May 29, 2016, http://www.newser.com/story/225645/average-size-of-us-homes-decade-by -decade.html.

23 Bill McKibben, *Deep Economy: The Wealth of Communities and the Durable Future* (New York: Henry Holt, 2007).

24 McKibben, *Deep Economy*, 108.

25 Eisenstein, *Sacred Economics*.

26 Eisenstein, *Sacred Economics*.

27 Frances Moore Lappé, Joseph Collins, and Peter Rosset, *World Hunger: Twelve Myths* (New York: Grove Press, 1998).

28 https://www.america.org/explore/research-publications/an-economy-for -the-1/.

29 Marc Ian Barasch, *Field Notes on the Compassionate Life: A Search for the Soul of Kindness* (Emmaus, PA: Rodale, 2005).

30 McKibben, *Deep Economy*.

31 Eisenstein, *Sacred Economics*.

32 Ed Diener and Martin Seligman, "Beyond Money: Toward an Economy of Well-Being," *Psychological Science in the Public Interest* 5, no. 1 (2004): 1–31.

33 David Myers, "What Is the Good Life?," *Yes!*, Summer 2004, 15.

34 Don Miguel Ruiz, *The Four Agreements: A Practical Guide to Personal Freedom* (San Rafael, CA: Amber-Allen Publishing, 1997).

35 Byron Katie, *Loving What Is* (New York: Three Rivers Press, 2002), 6–7.

36 Thomas Berry, *The Dream of the Earth* (San Francisco: Sierra Club Books, 1988), 171.

37 Ken Wilbur, *A Brief History of Everything* (Boston: Shambhala, 1996), 69.

38 Wilbur, *A Brief History of Everything*.

39 Katie, *Loving What Is*.

40 Peter Senge, C. Otto Scharmer, Joseph Jaworski, and Betty Flowers, *Presence: Human Purpose and the Field of the Future* (Cambridge, MA: Society for Organizational Learning, 2004), 27.

41 Inspired by "What-the-Bleep Study Guide," www.whatthebleep.com/guide.

42 Gregg Levoy, *Callings: Finding and Following an Authentic Life* (New York: Three Rivers Press, 1997), 139.

Chapter 8: Birthing a New Story

1 Ursula K. Le Guin, *Voices* (New York: Harcourt Brace, 2006), 166–67.

2 Katrina Shields, *In the Tiger's Mouth* (Gabriola Island, BC: New Society, 1994), 12.

3 Michael Dowd, *Thank God for Evolution: How the Marriage of Science and Religion Will Transform Your Life and Our World* (New York: Viking, 2008), 1.

4 "How Fast Does the Earth Spin," ScienceStruck, https://sciencestruck.com/ how-fast-does-earth-spin.

[5] David Abram, *Becoming Animal: An Earthly Cosmology* (New York: Pantheon, 2010).

[6] Abram, *Becoming Animal.*

[7] Richard Nelson, *The Island Within* (New York: Vintage, 1991), 249.

[8] Joanna Lauck, *The Voice of the Infinite in the Small: Re-Visioning the Insect-Human Connection* (Boston: Shambhala, 2002), 42.

[9] Daniel Goleman, *Social Intelligence* (New York: Bantam, 2006).

[10] Goleman, *Social Intelligence,* 43.

[11] Marc Barasch, *Field Notes on the Compassionate Life: A Search for the Soul of Kindness* (Emmaus, PA: Rodale, 2005).

[12] Leslie Brothers, "A Biological Perspective on Empathy," *American Journal of Psychiatry* 146, no. 1 (1989): 10–19.

[13] Michael Tomasello, *Why We Cooperate* (Cambridge, MA: MIT Press, 2009).

[14] Charles Eisenstein, *Sacred Economics: Money, Gift and Society in the Age of Transition* (Berkeley, CA: Evolver Editions, 2011).

[15] Angie O'Gorman, "Defense through Disarmament: Nonviolence and Personal Assault," in *The Universe Bends Toward Justice,* ed. Angie O'Gorman (Gabriola Island, BC: New Society, 1990), 242–46.

[16] Richard Rohr, *New Great Themes of Scripture* (Cincinnati, OH: St. Anthony Messenger Press, 1999).

[17] Sharif Abdullah, *Creating a World That Works for All* (San Francisco: Berrett-Koehler, 1999).

[18] Previous three paragraphs paraphrased from Uhl and Stuchul, *Teaching as if Life Matters.*

[19] Anthony DeMello, *Awareness: The Perils and Opportunities of Reality* (New York: Image Books, 1992).

[20] For assessments of America's recent history see Charles Eisenstein, *The Ascent of Humanity* (Harrisburg, PA: Panenthea Press, 2007); Thomas Berry, *The Great Work* (New York: Bell Tower, 1999); and Bill Plotkin, *Nature and the Human Soul: Cultivating Wholeness and Community in a Fragmented World* (Novato, CA: New World Library, 2008).

[21] "An Economy for the 99%," Oxfam Briefing Paper—Summary, January 2017, https://www.oxfamamerica.org/static/media/files/bp-economy-for-99-per cent-160117-summ-en.pdf.

[22] David Korten, *The Great Turning: From Empire to Earth Community* (Bloomfield, CT: Kumarian Press, 2007).

[23] Shannon Hayes, *Radical Homemakers: Reclaiming Domesticity from a Consumer Culture* (Richmondville, NY: Left to Write Press, 2010), 13.

[24] Robert Putnam, *Bowling Alone: The Collapse and Revival of American Community* (New York: Simon and Schuster, 2010).

[25] From a strictly economic perspective, it can be argued that time is money insofar as electing to spend one's time outside of the wage economy results in a loss of income—that is, there is an opportunity cost for withdrawing from the cash economy. But from the perspective of a person choosing to withdraw from the economy, this so-called opportunity cost has little, if any, relevance.

[26] Eisenstein, *Sacred Economics.*

[27] Schor, *Plentitude*; see also International Downshifting Week, "America: A Little Downshifting Never Hurt Anyone," March 29, 2011, http://downshiftingweek .wordpress.com/2011/03/29/america-a-little-downshifting-never-hurt-any one. Also see https://daily.jstor.org/the-road-to-utopia-a-conversation-with-ju liet-schor/.

[28] Schor, *Plentitude*, 101–2.

[29] Schor, *Plentitude*.

[30] Mike Barrett, "How Far Does Your Food Travel? 4 Reasons to Choose Local," *Natural Society*, December 17, 2013, https://naturalsociety.com/how-far-food -travel-food-miles-1500-average/

[31] And by cutting out middlemen, in the form of food processors, packagers, transporters, and retailers, farmers are justly compensated for their labor.

[32] The last CSA census was in 2012. At that time there were approximately thirteen thousand CSAs in North America; see USDA, National Agricultural Library, "Community Supported Agriculture," https://www.nal.usda.gov/afsic/commu nity-supported-agriculture. To locate CSAs and farmer's markets in your area go to www.localharvest.org.

[33] "Total Number of Farmers Markets in the United States from 1994 to 2014," Statista, https://www.statista.com/statistics/253243/total-number-of-farmers -markets-in-the-united-states/.

[34] Bill McKibben, *Eaarth* (New York: Henry Holt, 2010).

[35] "America's Electric Cooperatives: 2017 Fact Sheet," January 31, 2017, https:// www.electric.coop/electric-cooperative-fact-sheet/.

[36] Some of the food, transportation, and shelter options discussed here are also treated in my forthcoming book *Awaken 101* (Toplight, an imprint of McFarland).

[37] Fiza Pirani, "Why Are Americans So Lonely?," *Atlanta Journal-Constitution*, May 1, 2018, https://www.ajc.com/news/health-med-fit-science/why-are-americans -lonely-massive-study-finds-nearly-half-feels-alone-young-adults-most-all/bbIKs U2Rr3qZI8WlukHfpK/.

[38] Global Ecovillage Network, about page, https://ecovillage.org/about/about -gen/.

[39] Transition United States, "Official Transition Initiatives," http://transitionus. org/initiatives-map.

[40] Transition United States, "Transition 101," http://transitionus.org/transition -101.

[41] "First-Ever Zipcar Impact Report Shows Car Sharing's Significant Social and Environmental Benefits, and Its Essential Role in Creating Better Cities," January 15, 2019, https://www.zipcar.com/press/releases/impactreport2018.

[42] Rachel Botsman and Roo Rogers, *What's Mine Is Yours: The Rise of Collaborative Consumption* (New York: HarperCollins Publishers, 2010).

[43] Find Your Local Tool Lending Library, https://localtools.org/find/.

[44] Botsman and Rogers, *What's Mine Is Yours*, 163.

[45] Botsman and Rogers, *What's Mine Is Yours*, 124, 129.

[46] Paul Hawken, Amory Lovins, and Hunter Lovins, *Natural Capitalism: Creating the Next Industrial Revolution* (Boston: Little, Brown, 1999).

[47] Botsman and Rogers, *What's Mine Is Yours*, 130.

[48] Quoted at *Science and Philosophy*, http://sciphilos.info/docs_pages/docs_Einstein_fulltext_css.html.

[49] Marshall Rosenberg, *Nonviolent Communication: A Language of Compassion* (Encinitas, CA: PuddleDancer Press, 2001).

[50] Eisenstein, *Sacred Economics*, 3–4.

[51] Satish Kumar, *You Are Therefore I Am* (Dartington Totnes, Devon, UK: Greenbooks Ltd., Foxhole, 2001)

[52] Eisenstein, *Sacred Economics*.

[53] Eisenstein, *Sacred Economics*, 358.

[54] Puanani Burgess, "Blessings Revealed," *Yes!*, Winter 2009. A version of this story is also recounted in my forthcoming book, *Awaken 101* (Toplight, an imprint of McFarland).

[55] Wendell Berry, *Conversations with Wendell Berry*, ed. Morris Allen Grubbs (Jackson: University Press of Mississippi, 2007).

[56] Margaret Wheatley, *Leadership and the New Science* (San Francisco: Berrett-Koehler, 1999).

[57] Wheatley, *Leadership and the New Science*, 119.

[58] Mark A. Burch, *Stepping Lightly* (Gabriola Island, BC: New Society, 2000), 106.

[59] Wes Nisker, *Buddha's Nature* (New York: Bantam Books, 1998), 26.

[60] Nisker, *Buddha's Nature*, 26.

[61] Project for Public Spaces, www.pps.org/about/team/jwalljasper.

[62] Margaret Wheatley and Deborah Frieze, *Walk Out Walk On* (San Francisco: Berrett-Koehler Publishers, 2011), 58.

Epilogue

[1] Clarissa Pinkola Estes, "Inspiration," Aloha International, www.huna.org/html/cpestes.html.

[2] Susan Griffin, "Can Imagination Save Us?," *Utne Reader*, July–August 1996, 43–45.

[3] Charles Eisenstein, *Sacred Economics: Money, Gift and Society in the Age of Transition* (Berkeley, CA: Evolver Editions, 2011), 445.

[4] Pinkola Estes, "Inspiration."

Index

Page numbers in italics indicate figures.

Abram, David, 44, 53, 159
abundance, economism versus, 147–48
acceptance, 36, 130, 164
action: consequences of, 70; HPI and, 125; silence and, 97–98; social capital in, 166–69
ADD. *See* attention deficit disorder
adolescence, of culture, 15, 107, 181
advertising, 130
airscape, 53, 90, 127
algae, 77, 96
Amazon basin, 26, 69; cattle pastures in, 111; climate change and, 95
American Paradox, 148–49
ancestors, 16
anglerfish, 75
aphids, 30
applications. *See* practices
Arctic, 92–93, 96, 144
assumptions, questioning, 113, 127–29
astronomers, 6; stars and, 7
attention deficit disorder (ADD), 59
awakening: by turning inward, 175–76; by turning outward, 176–78
awareness, 176; of beliefs, 152–53; development and, 81–82; eating and, 60–61; expansion of, 21, *81*; pain and, 80; of universe embodiment, 5, 19

baboons, 83
Bacci, Ingrid, 78, 80
bacteria, 13, 32–33; cells of, 35; stars and, 14; surface area of, 47
Barasch, Marc, 161
bees, 30, 106–7
beetles, 30, 47; pine beetles, 95
behavior, 160

belonging: community and, 161; to Earth, 43–62; *Homo sapiens* and, 32; yearn for, 182
Berry, Thomas: on community, 23; on story, 17, 43, 151; on universe, 5
Berry, Wendell, 125, 174
BIA. *See* Bureau of Indian Affairs
Big Bang, 8–11, 16
Big Dipper, 5–6
big picture: belonging and, 58–59; courage and, 105–6; listening and, 78–80; Nisker on, 87; stepping back to see, 15–16, 36–37, 123–25
biochemistry, 25, 53
biodiversity, 57, 123
Biosphere, 33; unraveling of, 87–110; waste and, 122
Biosphere 2 experiment, 57–58
birds, 38, 66–69
birthing, 34; sites for universes, 10–11; women and, 113
Black, James F., 97
black holes, 10
blue jays, 26
Bohm, David, 107
Bormann, Herbert, 49
bottled water, 119
Bouman, Katie, 10
breastfeeding, 101
Buddhist monks, 160
Burch, Marc, 175
Bureau of Indian Affairs (BIA), 140
Burgess, Puanani, 173
butterflies, 25–29, 38, 178; migration of, 27–28, *28*
bycatch, 76
Byron, Katie, 152; *Loving What Is* by, 149